Dis Information

and other Wikkid myths

Dr KARL KRUSZELNICKI

ILLUSTRATED BY ADAM YAZXHI

HarperCollins*Publishers*

HarperCollins_Publishers_

First published in Australia in 2005
by HarperCollins_Publishers_ Australia Pty Limited
ABN 36 009 913 517
www.harpercollins.com.au

HarperCollins_Publishers_
25 Ryde Road, Pymble, Sydney NSW 2073, Australia
31 View Road, Glenfield, Auckland 10, New Zealand
77–85 Fulham Palace Road, London W6 8JB, United Kingdom
2 Bloor Street East, 20th floor, Toronto, Ontario M4W 1A8, Canada
10 East 53rd Street, New York NY 10022, USA

National Library of Australia Cataloguing-in-Publication data:

Kruszelnicki, Karl, 1948– .
   Dis information and other wikkid myths.
   ISBN 0 7322 8060 5.
   1. Science – Popular works. I. Title.
500

Cover photograph by Gerald Diel
Cover image art direction by Adam Yazxhi
Back image art direction by Adam Yazxhi and Caroline Pegram
Internal design and layout by Judi Rowe, Agave Creative Design
Bird's nest on p 114 drawn by Max Addison Yazxhi (aged 2½ years)
Printed and bound in Australia by Griffin Press on 80gsm Printspeed Offset

5 4 3 2 1   05 06 07 08

I dedicate this book to the Internet,
for its excellent role in disseminating
Dis Information …

Contents

Alcohol and Antibiotics

Antibiotics are one of the greatest success stories of modern medicine — up there with the discovery of vaccination, and the discovery that you shouldn't mix your drinking water and your toilet water. Like all drugs, antibiotics can have their bad side effects, but their benefits are enormous. Even so, some people wrongly believe that antibiotics are so dangerous that they should never be used. And, specifically, lots of people also wrongly believe that you should not drink any alcohol while taking antibiotics.

History of Antibiotics

Antibiotics go back a long way. The Chinese first used antibiotics about 2500 years ago. Back then they realised that the fungus that grew on soybean curd could cure boils. This ancient wisdom was known even earlier to the healers of Egypt and Mesopotamia. The soybean fungus was making a chemical (streptomycin), one of the first antibiotics. If you ate this antibiotic, it killed the bacteria that caused the boils. In fact, this same fungus today gives us streptomycin, which is our main defence against the bacterium *Yersinia pestis* that caused the bubonic plague.

In 1910, Paul Ehrlich helped introduce Salvarsan (containing arsenic) to successfully treat syphilis. The sulfonamide family of antibiotics was introduced in 1932, and some of them are still used today.

The first really powerful and widely used antibiotic, penicillin, was discovered by Alexander Fleming way back in 1928. Once again, it was made by a fungus. This fungus was called *Penicillium notatum*. Fleming noticed that a chemical made by this fungus would stop *Staphyloccus* bacteria from growing. Unfortunately he was a bacteriologist, not a biochemist, so he could not purify this mysterious chemical. Even so, he wrote about his discovery in the *British Journal of Experimental Pathology* in 1929.

In 1938 the brilliant biochemist Ernst Chain read this paper and managed to isolate and purify this mysterious chemical — not to try to invent a wonder drug, but just out of scientific curiosity. He called it 'penicillin', after its parent fungus. He worked in a pathology lab at Oxford University, which was run by Howard Florey. At first, Chain didn't get much support from Florey. Chain wanted to test his penicillin by infecting two mice with bacteria and then injecting penicillin to see if it cured them — but he didn't know how to do injections, and Florey wasn't interested in helping. Chain had to get another colleague to inject the mice first with the bacteria and then with his mysterious penicillin. The penicillin worked, the mice recovered fully — and suddenly Florey was very interested.

The Golden Age of Antibiotics

The next step was to test pencillin on humans. It was painstakingly difficult to get any penicillin at all, but eventually they had enough. In 1940 a 48-year-old London policeman, Albert Alexander, made a tiny cut in his skin while shaving. A bacterium invaded his body through this cut and soon infected him. At death's door, he was rushed to the Radcliffe Hospital with a temperature of 40.5°C. Florey and Chain gave him penicillin and he began to recover. After five days they ran out of penicillin, and

The Loving Lurgies

It's long been thought that the consumption of alcohol affects the performance of antibiotics.

Antibiotics

Alcohol

While it's never a good idea to drink alcohol while on medication, this story really started in the '50s when antibiotics were used to combat STDs. When prescribed to afflicted folk, they were given the sombre advice that alcohol should never be used while taking penicillin.

The real reason for this advice was to stop STD-carrying folk getting drunk and frisky and spreading their 'infected love' around the town.

I'm drunk ... itchy, medicated and I need some lovin'

Alexander began to worsen. They managed to purify some penicillin out of his urine and he began to recover again. Finally there was no more penicillin, and Albert Alexander died. Florey and Chain had showed that penicillin worked, and that it wasn't harmful. So the treatment worked — but the patient died.

Tiny quantities of penicillin were soon made available to cure a very small number of people. Even though Fleming had long stopped researching *Penicillium notatum*, he still kept an eye on it. He had a friend who was dying from an infection, and managed to get him cured with some of the precious penicillin.

Chain and Florey were very concerned that a Nazi air raid might destroy their building and all their research. Before they went home each night, they would rub some of the raw fungus inside the pockets of their trousers. So if their lab was flattened overnight, they would still have some fungus left with which to start up again.

More is Better

Chain had been brilliant in isolating and purifying penicillin; Florey was brilliant in the next stage — ramping up from small quantities to mass production, by using some of the technology employed in brewing beer. He grew the *Penicillium notatum* fungus in enormous 95 000-litre vats with air bubbling through. This meant that the fungus could grow anywhere in the volume of the tank, not just on the surface.

But *Penicillium notatum* did not make enough penicillin, especially of the quantity required to deal with the massive number of wounds and infections caused by World War II. As soon as the scientists realised this, they set out on a search for more productive penicillin moulds, from rainforests to the backs of people's fridges. Mary Hunt, a laboratory worker in Peoria, Illinois, hit the jackpot when she brought in a canteloupe (rockmelon) infected with a 'pretty, golden mould'. This particular fungus was *Penicillium chrysogeum* — and, in its natural form, it delivered 200 times as much penicillin as *Penicillium notatum*. By the time Florey and his American colleagues had finished mutating the new fungus with X-rays, and selectively breeding it, they had something that yielded 1000 times more penicillin than *Penicillium notatum*.

In the five months from January to May 1943, they could make 400 million units of penicillin. By the end of 1945, the American factories were making 650 billion units each month. Also in 1945, Fleming, Chain and Florey shared the Nobel Prize for Physiology or Medicine.

Wonder Drug

The first batches of penicillin in 1943 were reserved specifically for the military. Later, as larger quantities were produced, it was made available to civilians, first only for life-and-death cases, and then for general use in the community.

Penicillin was truly a miracle drug when it was first introduced. It worked quickly and effectively against pneumonia, meningitis and hundreds of other deadly diseases. It was also especially effective against what were then called venereal diseases (VDs) and are now called sexually transmitted diseases (STDs).

The Myth

The VD clinics of the 1950s and '60s gave the sombre and serious advice that alcohol should absolutely never be used while taking penicillin. But, in reality, there were — and are — no significant chemical interactions between penicillin and alcohol. The real reason the VD doctors and nurses gave this advice was based on moral, not pharmacological, grounds. They were worried that alcohol would reduce the inhibitions of the VD sufferers, who, while under the influence, might get a little 'frisky' and pass on their infections to other people — before the penicillin had a chance to cure the sexually transmitted diseases.

And that's how the mythconception that alcohol should never be taken with antibiotics arose.

Even so, it's well known that alcohol can interact quite nastily with a small number of modern drugs such as tinidazole (brand name Fasigyn) and metronidazole (Flagyl), potentially causing nausea, vomiting, abdominal cramps, headaches, fast heart rate and flushing. And alcohol can reduce the absorption of other antibiotics, such as the doxycyclines and some other tetracyclines. But these few interactions are well known to both medical doctors and pharmacists.

Mind you, too much alcohol can put an extra load on your liver and immune system, can impair your judgment, liberate aggressive tendencies, reduce your energy state — and be associated with staying up late, behaving recklessly and not getting all the rest that your body needs to heal itself. So half a glass of an alcoholic beverage of your choice would be fine with most antibiotics, but probably no more than that …

Another True Reason

Another 'reason' people were told about why antibiotics and alcohol did not mix was to help keep the peace while at war. The story goes (and I have no written reference for this, only personal anecdotes from World War II soldiers) that during World War Two there was some occasional friction between the American and British military units stationed in Great Britain. Apparently, there were so many fist fights while under the influence of alcohol that the US Army Surgeon General recommended that the soldiers should not mix alcohol and antibiotics.

Penicillin Kills Guinea Pigs, but Not People

Most drugs are tested on animals before they're released onto the market. Luckily for us, penicillin did not get tested on guinea pigs — because it kills them. If penicillin had been developed in peacetime, and in a more litigious society, it might have never come to market.

References

Carr, Nick, 'Medical Myths' www.abc.net.au/dimensionshealth/transcripts/s776185.htm

Fasigyn, Flagyl, MIMS Annual 2006, Published by MIMS Australia, ISSN 0725-4709, pp 8–802, 803.

Encyclopaedia Britannica, Ultimate Reference Suite DVD, 2005.

Jetsetting Germs

I spend a lot of time inside those high-speed aluminium tubes with wings called jet planes. The maximum number of flights I've had in one calendar year is 104, but nowadays I'm more sensible, so last year I took only 67 flights. Many people believe that in planes the air conditioning system spreads the germs through the whole plane, and so you are more likely to pick up an infectious disease from other passengers in the plane — but that's wrong.

Going on a holiday, leaving your daily life behind and arriving somewhere lovely and different, is very nice. The actual experience of flying, however, is not that pleasant. For example, an aircraft cabin gives the smallest volume of air per person of any public space. So, while the air conditioning may not give you diseases, it doesn't really give you that much air.

Thrill of Flying

Your typical jet plane is noisy. When I started flying a lot, I noticed that my fellow passengers coming off jets were talking to each other very loudly. This was a result, I thought, of them being slightly (and temporarily) deafened by the background noise on the plane — from the air rushing over the body, the jet engines, and so on. So I started wearing earplugs during an entire flight. When I took them off at the end of a flight, I noticed that the other passengers were talking to each other more loudly than they had at the beginning of the flight, which seemed to prove my theory. Nowadays, I wear noise-cancellation headphones on jets with plug-in sound systems, and earplugs on puddle-jumpers (planes with propellers).

Your average jet plane is very dry. The air is at 10–20% humidity instead of a more pleasant 40%. Each person on a plane breathes and sweats out about 100 ml of water each hour. On a typical trans-Pacific flight (for example, Sydney to Los Angeles), this works out to be approximately 600 kg of water. (By the way, some of this will condense in the cavity between the cabin wall and the aluminium skin of the plane and can possibly cause corrosion, or degrade the wiring insulation.)

Fasten seat belts: and wipe your nose

It is commonly believed that a plane's air-con system spreads germs through the whole plane — but that's wrong.

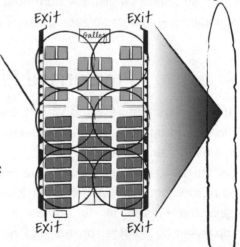

A plane's air-con system is divided into blocks of about 5 rows. The air tends to stay in its little area, and not flow along the length of the plane.

But if a sick person walks along an aisle, coughing all the way, they could very well spread germs right down the plane.

Exit   Exit

Galley

Exit   Exit

Typical jet-aircraft seating plan

The air pressure in the cabin is lower than normal — about 70–80% of sea-level air pressure. You breathe in less volume of air with each breath, so your oxygen saturation levels are lower. In one study of 84 people aged 1–78 years, about half had their oxygen saturation level drop to a relatively low 94% when the plane was at cruising altitude. If you were a patient in a hospital with an oxygen saturation level of 94%, the hospital doctors would seriously consider giving you oxygen from a mask. This low oxygen level would slightly reduce your thinking capacity; the low air pressure is also thought to cause your legs to bloat a little.

Finally, planes are squashy. Dr M.B. Hocking from the Department of Chemistry at the University of Victoria in Canada wrote that 'air travellers represent one of the most diverse groups of people called upon to sit in close proximity for prolonged periods with the smallest available air space per person of any current social setting'. Each passenger in an aircraft cabin gets 1–2 m^3 of air space. This is much less than the 4–10 m^3 of air they would get in an auditorium, 10–20 in a ballroom, or 10–30 in an office building.

Because of these cramped conditions, many people believe germs run rampant on planes and leap happily from one person to the next in a domino effect. They also believe that they are likely to pick up an infectious disease from other passengers because the air-conditioning system spreads bacteria through the entire plane.

Pressurised Planes

In the old days (before World War II), planes were not airtight and air leaked in and out. The air pressure inside the plane was the same as the air pressure outside.

In the 1940s, aircraft cabins were pressurised so that the crew (and passengers) would remain conscious at higher altitudes. In a jet engine the outside air is sucked in, heated (to 250°C) and compressed (to 30 atmospheres — 1 atmosphere is 10 tonnes per m^2). These conditions will definitely kill all known germs. Some of this sterile, hot, compressed air is bled off, expanded, cooled down and then fed into the cabin to increase the pressure. Modern planes have several outlet valves through which the cabin air 'leaks' into the thinner air outside. The flight crew in the cockpit controls the cabin pressure by operating these valves — how many are open, how far open they are, etc.

There are advantages and disadvantages to a pressurised cabin. On one hand, a plane flying at 30 000 ft (9 km) saves about 38% on fuel costs. This is because the air is thinner at this altitude, so there is less resistance to the plane pushing through the air. But on the other hand, the plane has to be pressurised so that the passengers and crew stay conscious. Thus, some of the 'working air' has to be stolen from the engine to pressurise the cabin. This gives a penalty of 2% of fuel cost for the extra energy needed to pressurise the cabin. You can see that the fuel savings (38%) massively outweigh the costs (2% extra).

Cabin Air — Dump or Recycle?

Back in the early days of pressurising cabins the air was fed in and then dumped, so that the passengers breathed the air only once. This was very good for the passengers, but it cost fuel. So the plane-makers introduced recycling of some of the cabin air, in order to bleed less air from the engines. The 1945 Boeing Stratocruiser was one of the first planes to recycle passenger cabin air.

The advantage of recycling air is that it saves the airlines about US$60 000 per jet plane each year. The disadvantage is that the passengers breathe air with higher than normal levels of carbon dioxide. Other factors (such as low humidity) make the air not quite as lovely as rainforest air.

The fact that the flight crew determines how much of the air is recycled and, hence, the quality of the air you breathe helps explain why sometimes you come off a plane feeling okay, while other times you might have a slight headache. By the way, the flight crew in the cockpit usually has totally fresh air that has not been recirculated, for safety reasons.

Back in 1970, a passenger in an average aircraft of the day would get 7 litres per second of outside air. By 2000 this had dropped as low as 2.8 l/sec and, occasionally, 1 l/sec. You can see this trend in different models of the Boeing 737: a 1967 737-100 provided 7.6 l/sec, but this had dropped to 4.8 l/sec in the 1984 737-300. This amount of outside air is much less than that recommended by many experts, who would prefer the rate to be 15 l/sec.

Aircraft Air and Germs

Today, recycled air in planes is cleaned with filters that can remove bacteria. These filters are present in about 85% of the planes that carry over 100 passengers, but usually not in smaller planes.

So what about germs swirling around in the cabin air of a jet with air filters? It turns out that you are probably more protected from germs in a plane than if you work in an office.

The plane's air-con system is divided into blocks of about five rows of seats. The air tends to stay in its little block, and not flow along the length of the plane. The air in the plane's cabin comes out of vents in the ceiling, flows over the passengers and then back into the air-con system via vents at floor level. Each cycle takes about 3–5 minutes. About 63% of germs are removed on each cycle, thanks to the combination of some fresh air being admitted on each cycle, and the filters. So the germs from one person are not spread along the length of the plane by the air con — they stick in the block of five seats. You can be infected by the air con on a plane, but only by people near you — which is the same as going to the supermarket, or travelling on a bus. However, if a sick person walks along the aisle coughing all the way, they might spread germs right down the plane.

Therefore, it's not impossible to get sick on a plane. And one study predicted that the infection rate on a plane could be halved if you doubled the ventilation rate. But this would cost about $1 in fuel per passenger on an eight-hour flight, and in that case most airlines would earn less on their investment than they would from putting the monetary equivalent to that investment in a bank and earning interest. On the other hand, how much does it cost society to care for sick people? Maybe the airlines could crank up the ventilation rate during outbreaks of infectious diseases?

Dr Ron Behrens, a consultant in Travel Medicine at the London School of Hygiene and Tropical Medicine, said in *New Scientist* magazine that departure lounges at airports are far riskier than the actual flight.

Aircraft Carry People with Germs

There is one major infectious diseases risk associated with planes — spreading diseases around the world.

Each year, over one billion passengers fly on planes, with about 50 million passenger flights to the poorer countries (where many new diseases start). Since the 1960s, influenza epidemics have followed major air-transport routes. Back in the 14th century, the Black Death took about three years to get from southern Italy to Great Britain via rats. Today, a passenger on a jet can carry infectious diseases anywhere in the world in a few hours — but probably won't infect the other passengers on the jet.

Squashy Planes

Over the years the airlines have reduced the seat pitch (distance between the seats) in an effort to stuff in more passengers. The SeatGuru website (**www.SeatGuru.com**) gives information regarding seating on specific aircraft on specific airlines, e.g. which seats on the plane are extra quiet, have extra leg room or a non-restricted recline.

Spreading Disease with the Greatest of Ease

Different diseases range in how infectious they are. They can be very infective, such as the common cold, or much less infective, e.g. leprosy.

They spread by different methods too. Some will spread only by direct blood-to-blood contact; others will spread by kissing or prolonged face-to-face contact with one person breathing in another person's exhaled air. Other diseases will spread when one person inhales the infected saliva droplets of an infected person.

Different diseases also need different numbers of infective particles (bacteria, viruses, etc.) to have a reasonable chance of infecting you.

It is not a good survival policy for a disease to be very nasty and very quick to act. In that case, it might kill its host before it has a chance to infect another person, and then it might die out.

UV Light and Sick Buildings?

'Sick Building Syndrome' was first recognised in the 1980s. People in such buildings suffer respiratory symptoms such as asthma, itchy eyes, stuffy nose, and so on. It turns out that some of these symptoms are caused by germs (such as bacteria and fungi) growing in various parts of the air-conditioning system.

Ultraviolet light has lots of energy. It some cases, it has enough energy to kill germs. Dick Menzies at McGill University in Montreal, Canada, installed some UV Germicidal Irradiation lamps in the air-conditioning systems of three office blocks. This treatment reduced people's symptoms by 20–40%. The effect was greatest in non-smokers and people with allergies.

The treatment was certainly cost effective. It cost $52 to install and $14 to run for a year. It would pay for itself in less than six months if it reduced each worker's absenteeism by one day per year.

Perhaps this treatment could be used in planes, to make them even safer?

References

Bhattacharya, Shaoni, 'Ultraviolet light can cure "sick buildings"', *New Scientist*, 28 November 2003.

Coghlan, Andy, 'Boosting aircraft ventilation may cut disease', *New Scientist*, 11 March 2005.

Hocking, M.B., 'Passenger aircraft cabin air quality: trends, effects, societal costs, proposals', *Chemosphere*, 41, 2000, pp 603–615.

Humphreys, S., et. al., 'The effect of high altitude commercial air travel on oxygen saturation', *Anaesthesia*, 2005, Vol 60, pp 458–460.

Ozonoff, David and Pepper, Lewis, 'Ticket to ride: spreading germs a mile high', *The Lancet*, 12 March 2005, Ozonoff.

Bullets Fired Up

One of my readers asked me this very sensible question: 'We often see overseas images of crowds of people firing off hundreds of rounds of bullets into the air in celebration. What happens to those bullets? Surely they cannot keep heading into space for ever? When gravity finally takes over, why aren't they falling back and killing the same people that fired them off in the first place?' I've never been in one of these crowds, but I would guess that most of these revellers think that the falling bullets can't hurt them or their friends. They are very wrong.

Going Up

The first thing to realise is that what goes up usually comes down. In the case of a bullet fired upwards, it *will* come down — but more slowly.

In most of these celebrations, the bullets are fired from military rifles. A rifle bullet is fired with a typical muzzle velocity of around 2700 ft per second, or around 3000 kph. Once the gases stop pushing it, and it has well and truly left the barrel of the gun behind, it begins to slow down.

Two forces are acting together to slow it down. First, the resistance of the air that it's pushing through slows down the bullet.

Second, the downward suck of gravity also slows the bullet. On average, a bullet will take around 20 seconds to climb to a height of around 3 km, at which point it has come to a dead halt. Then, after the briefest of instants, it begins to fall towards the ground.

So far we have looked at the bullet slowing down from 3000 kph to zero. Now let's look at the bullet accelerating from zero as it falls.

Coming Down

As the bullet falls, once again it's subject to two forces. The big difference is that this time the forces are acting against each other. On one hand, the suck of gravity is trying to make the bullet fall faster. But on the other hand, the wind resistance is trying to slow down the falling bullet.

The suck of gravity is not as powerful as the exploding gases that pushed the bullet out of the barrel. So the plummeting bullet will not accelerate to a maximum speed of 3000 kph; instead, its top speed is somewhere between 330–770 kph, depending upon the weight and shape of the bullet. Because it's falling more slowly on the way down, it takes about 30 seconds to reach the ground from the top of its flight.

This is the essential difference: the maximum speed on the way up is around 3000 kph, but on the way down it's about 770 kph.

Injuries and Death

A speed of 770 kph is not as fast as 3000 kph — but it's more than enough to penetrate any human skull — you need a velocity of only around 220 kph to do that (110 kph will penetrate skin while 220 kph will shatter bone, according to studies by the American National Rifle Association). Most people who have been hit by bullets falling out of the sky get hit on their upper back, the top of their head or their shoulders.

In Kuwait after the end of the 1991 Gulf War, the Kuwaitis celebrated by firing weapons into the air — and 20 Kuwaitis were killed by falling bullets. In Los Angeles between the years 1985 and 1992, doctors at the King/Drew Medical Center treated some 118 people for random falling-bullet injuries; 38 of them died. According to the *Los Angeles Times*, most of the injuries happened during 'raucous Independence Day and New Year's Eve celebrations'. Most of the survivors were left with severe long-term disabilities such as chronic pain, seizures, paraplegia and quadriplegia. Every single one of them was taken by surprise — they did not see anybody with a gun, and they did not hear any gunfire.

Unfortunately, this tradition of firing bullets into the air for celebrations has arrived in Australia. A nine-year-old girl in Belfield, Sydney, was watching fireworks with her parents from their driveway to celebrate the incoming New Year of 2002. At 12.05 a.m., a bullet fell out of the sky and lodged in her upper arm. If she had been standing a few centimetres to one side, she could have been killed ...

What goes up: can also kill you

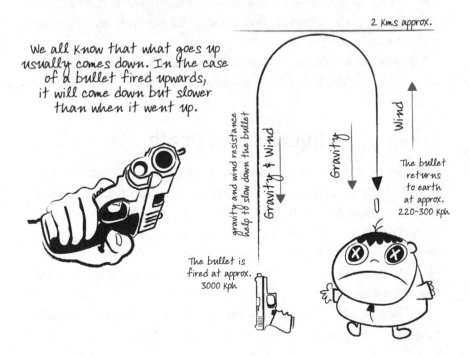

2 Kms approx.

We all know that what goes up usually comes down. In the case of a bullet fired upwards, it will come down but slower than when it went up.

gravity and wind resistance help to slow down the bullet

Gravity & Wind

Gravity

Wind

The bullet returns to earth at approx. 220–300 Kph

The bullet is fired at approx. 3000 Kph

USA Experience

In Slidell, Louisiana, in the USA, it is illegal to shoot firearms into the air to celebrate (say) the New Year. The maximum penalties are six months in jail and a $500 fine.

In New Orleans, Louisiana, the falling bullets problem is so severe that there is now a 'Falling Bullets Kill' ad campaign between Christmas and New Year. It began in the mid-1990s after a tourist was killed by a falling bullet on the riverbank in the middle of the French Quarter. The more cautious folk usually head indoors (to be under a thick roof) for about a quarter of an hour each side of midnight.

According to the National Institute of Justice, the research and development agency of the US Department of Justice, people have been killed by falling bullets in New Orleans, Detroit, Kansas City, Philadelphia, California and Missouri.

USA Experiment

There have been very few experiments to actually measure the velocity of falling bullets. The US Army Ordnance conducted one of the first in 1920.

They set up a 3 m x 3 m platform in the middle of a lake or inlet near Miami, Florida. The platform was covered with a thin sheet of armour plate, to protect the soldiers. The gun fired .30 calibre, 150-grain. Spitzer point bullets at 2700 feet per second (3000 kph) muzzle velocity. The gun's orientation could be adjusted to bring the falling bullets near the platform.

The soldiers fired a total of 500 bullets. Four landed on the platform, one on their boat moored to the platform, some on land, and the rest in the lake (making an audible splashing sound). The bullets from each burst of firing would land over an area about 25 m across.

Wind had a dramatic effect on where the bullets landed. An 8 kph ground wind would have the bullets landing 100 m away. But the winds aloft are very variable, and could even blow in a completely different direction.

ShotSpotter

A new technology, ShotSpotter, uses microphones and earthquake detection mathematics to find just where the bullets were fired from. There are claims that the accuracy is to within 8 m.

One early test was done with this technology for the New Year of 1999/2000, in the 2.5 km² Willowbrook area of Los Angeles. In just two hours it detected over 1200 'celebratory' gunshots. The police at the station knew the location of the firearms within 15 seconds of their being fired.

Bullet on Moon

The situation is different when the bullet is fired upward through a vacuum, such as on the surface of the Moon, as there is no wind resistance. So, if the bullet leaves the barrel at 3000 kph, it will hit the ground at the same speed — 3000 kph. It will also climb much higher (35 km) and take 168 seconds for the round trip.

References

'Are Bullets Fired Into the Air Lethal?', *Focus*, November 1993, p 38.

Kennedy, Les, 'Girl Shot In Arm As She Watched Fireworks', *Sydney Morning Herald*, 11 January 2002, p 3.

Maugh, T.H., 'Bullets Fired at Sky Cited in 38 Deaths', *Los Angeles Times*, 30 June 1995, Section B1.

Roylance, Frank, D., 'Police Sound Out Gunfire's Source', *Baltimore Sun*, 7 April 2003.

Attack of the Killer Piranhas

Our shallow Western culture loves cheap thrills, and so we embrace the gruesome reputation of the piranha fish. According to all the authoritative sources (such as James Bond movies, the movie *Piranha II* and the console game *Tomb Raider III*), piranhas can strip all the flesh off a man within minutes. The mental image is that you change from a fully fleshed human being to a cartoon skeleton almost immediately. Even former US President Teddy Roosevelt wrote after a trip to Brazil in 1913 that the piranhas are 'the most ferocious fish in the world. They will snap a finger off a hand incautiously trailed in the water. They will rend and devour alive any wounded man or beast'. But it seems that the piranhas are the mostly innocent victims of a cruel mythconception.

40 Species, Mostly Mild

There are about 40 species of piranha, found in South America from north Argentina to Colombia, but only about three species (especially the Red-bellied Piranha) show any real signs of aggression. In the local Amazonian Tupi language, the word '*piranha*' means 'toothed fish'. Some languages or dialects give them the name '*caribe*', which means 'cannibal' or, sometimes, 'donkey castrator'.

Piranhas range in length from a few centimetres up to 60 cm. They do indeed have fearsome, pointed, triangular teeth. A few of the species are vegetarian, but most of them eat meat. Most of the species of meat-eating piranha will simply graze on other fish, taking a single, small, semicircular bite out of a fin or scale before letting the wounded fish swim away. Each fin or scale is between 35–85% protein, making it very nutritious. In addition, the blackwater rivers of South America are often low in calcium and phosphorus, and fish scales or fins are loaded with these minerals. The fin or scale of the victim will grow back in a few weeks so the piranha can have another feed. Piranhas are like cattle that gently nibble on only some of the grass in a sustainable fashion.

A few of the piranha species are almost totally vegetarian, preferring to feed on flowers, fruit, nuts and seeds. This diet

Bite me!

According to the authoritative sources (James Bond movies, Lara Croft games and the movie Piranha II) piranhas can strip all the flesh off a man within minutes.

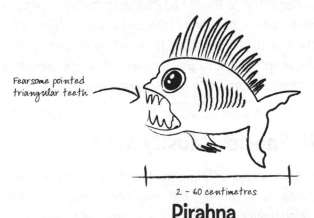

Fearsome pointed triangular teeth

2 – 60 centimetres

Pirahna
(Amazon Tupi language for 'Toothed Fish')

probably evolved with the massive Amazon floods that can inundate an area the size of England for seven months at a time. Some 200 species of fish, including a few species of piranha, migrate into this new feeding ground to eat and make babies. In many cases, the seeds need to pass through the piranha's gut to better help them germinate.

Fishy Story

In general, the carnivorous piranhas prefer to graze or scavenge off dead animals rather than attack a whole healthy animal. They also tend to travel in schools of 20 or more; even so, they are usually quite timid.

The Brazilian scientist Professor Ivan Sazima from the Institute of Biology, Universidade Estadual de Campinasin, has studied piranhas. In many years of research, his team could not find one single case of piranhas killing a human. Instead, they found only people who were already dead in the water, *before* the piranhas came along to have a nibble. For example, one person's drowned body had indeed been reduced to bones by losing all its flesh to piranhas — but that was after four days in the river. After 20 hours in the water, another person's drowned body lost the flesh off the arms and legs, but not the trunk. A third person, who had died from a heart attack while in the water, lost only small amounts of flesh after a few hours in the water.

In no case was it like in the James Bond movies, in which the unwanted villain has every shred of flesh torn from him as he vanishes into the foaming water. There has never been found a case similar to Teddy Roosevelt's description of a man travelling on a mule which returned to camp without him. Roosevelt wrote that when his travelling companions found the skeleton in the water, 'his clothes [were] uninjured, but every particle of flesh stripped from his bones'.

'Boiling' Water

However, there are a few rare cases where the piranhas (usually the red-bellied ones) will actually 'make the water boil'. But these are very special cases.

One situation is when the local fishermen will deliberately throw unwanted fish guts into one small part of the river. Over time the piranhas learn to migrate to where the eating is good, and will quickly demolish (for example) a plucked dead chicken if you throw one into the water. These piranhas have been conditioned to eat bleeding guts, and will go into a feeding frenzy over a chicken. But just a few hundred metres upstream, a different bunch of piranhas of the same species, who have not been so conditioned over a period of time, will not get as excited over a plucked dead chicken.

Another situation in which the water 'boils' with voracious piranhas is when a whole bunch of birds have their nursery in a tree that overhangs the river. The parents nourish their babies by vomiting up food for them. Their aim is not perfect, and the babies don't yet have very quick and accurate reflexes, so some of the regurgitated food drops out of their mouths into the river — and the piranhas learn to hang around for this Free Food From Above, just as they would for the chicken. Once they have been conditioned to eat this Free Falling Food, the piranhas will attack any baby bird that drops into the river below.

A circumstance in which you should be cautious is when you have just caught a piranha which is now flapping about in the bottom of your boat. This piranha is quite annoyed, and it's probably better to keep away from it. This, however, is not the 'water-boiling' situation.

The Truth

But in general, apart from these few specific cases, piranhas are harmless to humans, who can happily swim in the same water as the fish. In the July 1999 issue of *The Smithsonian*, Richard Conniff describes how he swam with piranhas, both in a tank in Dallas and in the Amazon rainforests, and came to no harm. While he did write that he wouldn't try skinny dipping with them, he also noted that the Amazonian kids would — and never came to any harm.

In general, the vast majority of piranhas 'lead lives of quiet desperation. Instead of swarming over their victims in a tumult of flashing teeth, piranhas mostly lurked and stalked and even disguised themselves as other species to snatch their food on the sly,' according to Conniff. Thus, with the piranha, their bark is worse than their bite.

More Dams, More Piranhas

In late 2003, Professor Ivan Sazima wrote about how increased damming of rivers had led to increased piranha attacks on bathers. He described in the scientific journal *Wilderness and Environmental Medicine* how parent piranhas looking after their babies would protect them from humans — with just one bite per human.

A river is commonly dammed to slow down the water so tourists and locals can go swimming, or for flood protection as the population increases. But the speckled piranha (*Serrasalmus spilopleura*) likes to breed in slowly flowing water. It lays its larvae in floating or submerged waterweed (e.g., water hyacinth). Rapidly flowing water usually washes these plants away, but they survive better in still water, so piranha numbers can increase tenfold in such conditions. The fretful piranha parents guard their larvae, who swim comfortably in the shelter of these plants.

Unfortunately, swimmers can accidentally disturb these plants — and the baby pirahnas. The proud piranha parents swoop in to give a single warning bite to the swimmer, leaving behind a circular bleeding wound. But the other piranhas would leave the swimmer alone and never do the rip-flesh-to-the-bones sequence for which they are infamous in the movies.

The town of Santa Cruz of Conceicao, on the Rio Mogi Guacu river, never had any piranha bites before the river was dammed in 1998. But, since then, the incidence of piranha attacks has increased. Over the short period of five weekends, there were 38 piranha attacks — all to defend piranha babies.

Piranha — the Original Funding Fraud

Barry Chernoff, the Professor of Environmental Studies at Wesleyan University, calls the piranha 'the fake monster'.

Perhaps only half joking, he reckons that the original European explorers of South America had to return with specimens and stories to impress the fund-givers enough that they would give the explorers another batch of money to return and continue their explorations. He says that piranhas were 'probably the original funding experiment'.

References

'In His Own Words: Barry Chernoff, Professor of Environmental Studies at Wesleyan University', *Discover*, July 2004, pp 58–65, as told to Cal Fussman.

Conniff, Richard, 'Relax, It's Only a Piranha', *The Smithsonian*, July 1999, pp 42–50.

Haddad, Vidal and Sazima, Ivan, 'Piranha Attacks on Humans in Southeast Brazil: Epidemiology, Natural History, and Clinical Treatment, With Description of a Bite Outbreak',*Wilderness and Environmental Medicine*, Vol 14, No. 4, pp 249–254.

Alien Autopsy — the Roswell Case

There's an old saying in medicine: 'You can always make a correct diagnosis — but sometimes you have to wait until the autopsy!' The autopsy is the gold Standard of Credibility — it's the only test that reveals what really killed a patient. So when the Fox Network in the United States showed film of the autopsy of an actual alien body recovered after the Roswell Incident of 1947, there was huge interest. Maybe this film would finally prove that intelligent aliens have been visiting our planet.

According to a 1997 Gallup Poll, about 80% of Americans had heard of the 'Roswell Incident', and about 30% thought that it involved an alien spaceship crashing at Roswell, New Mexico. In a nutshell, the Roswell Incident was the supposed recovery of strange debris and a few dead aliens from a crashed alien UFO (Unidentified Flying Object), and the subsequent cover-up by the military.

Important Background Info

First, the Roswell Incident happened only a few weeks after an Idaho businessman and pilot, Kenneth Arnold, claimed that he had seen UFOs around 3 pm on 24 June 1947, while flying near Mt Ranier in Washington State. He said that he saw nine silvery disc-like objects darting about in the sky and described their agile movements as being 'like pie plates skipping over the water'. It was the newspaper reporter who came up with the phrase 'flying saucers', which has stuck ever since. This report started up a near hysterical short-lived flurry of UFO sightings, which then quietened down to the steady trickle of reports that we get today. But in early July 1947, UFO sightings were everywhere.

Second, Roswell Army Air Force Base was home to the 509th Bomber Wing. The USA had used up three of its atom bombs by the end of World War II — one was tested in New Mexico, the other two were dropped on Japan. At the time, the USA successfully bluffed the rest of the world into thinking that they still had lots more left. By the time of the Roswell Incident, the USA had manufactured only a handful of nukes, and the 509th Bomber Wing at Roswell was the only bomber wing on the planet that carried atom bombs — so the base was under very high security.

Third, both parties involved (the military of the day and, later, the conspiracy theorists) have told tiny lies, little lies and big lies; also, they have accidentally said things they didn't mean to say, as well as deliberately not said things that they knew to be true and that they should really have told us. They also changed their stories over time. In other words, they were just like the rest of us. So it's been very difficult to get close to what actually happened. But this is probably pretty near the truth . . .

Roswell — Project Mogul

In 1947, the USA launched several special top-secret high-technology balloons as part of Project Mogul. These balloons were designed to go to a set height and remain there, listening for the sounds of possible Soviet nuclear tests. During World War II the US military had mathematically analysed the Krakatoa volcanic eruption of 27 August 1883, which was so loud that the sound was heard in Australia, some 3500 km away. The US military scientists had worked out that there must be some kind of 'sound channel' or 'acoustic duct' in the upper atmosphere that caught and confined the sound of the Krakatoa eruption, and then carried it around the world. So perhaps if they could place listening balloons at the right height, they could hear sounds from the other side of the planet — and they hoped that these sounds would include clandestine nuclear blasts.

Each balloon was not a single balloon. Instead, it was actually a series of smaller balloons, batteries, radar reflectors, listening devices and so on, strung vertically together, hanging downwards for over a hundred metres.

As it turned out, however, the Soviets didn't explode their first nuke until two years later, in 1949. Also, the Project Mogul balloons were never really successful as listening devices. But the project did develop techniques that were later very useful in other projects; these included using planes to catch payloads dropped from balloons, which later evolved into using planes to catch payloads of undeveloped film dropped from spy satellites — and the highest ever jump from a balloon (Joe Kittinger, 1960, 31.3 km).

Roswell — The Incident

In 1995, one of the three surviving scientists who had worked on Project Mogul, Professor Charles B. Moore, openly spoke about it. As a cover, some of the flights had been officially known as New York University (NYU) balloon flights. Moore reckons that the strange debris of the Roswell Incident came from NYU Flight #4, which he personally helped launch on 4 June 1947, from Alamogordo Army Air Field. NYU Flight #4 was tracked until its batteries ran flat. By then, the winds had carried it north-ish to about 27 km from where the Roswell debris was finally found.

Alien autopsy

The Roswell incident was the supposed recovery of strange debris and a few dead aliens from a crashed UFO – and the subsequent cover-up by the military.

There were a few crucial mistakes that appeared in the autopsy footage:

1. The alien's body 'sagged' in the wrong direction.

2. The arms and legs never bent past the angle at which synthetic materials would usually buckle.

3. The alien's buttocks were firm ... dead animals lose all muscle tone.

On 14 June 1947, William W. 'Mac' Brazel, a sheepherder foreman on the Foster Ranch, found some very strange debris strewn across one of the ranch's fields. The debris was made up of shiny thin foil, grey rubbery material, small sticks and heavy paper, amongst other things. Brazel reported his findings to the local sheriff, and soon the nearby Roswell Army Air Field heard about it. An overenthusiastic report to the newspapers by Military Intelligence at Roswell spoke of a UFO, but this was retracted the next day; conspiracy theorists made claims of UFOs and aliens confiscated by the US Air Force. But, without any hard evidence, the mainstream media soon lost interest, with only dedicated UFO-logists still believing.

Roswell — The Rush

The Roswell Incident was mostly forgotten until 1980. That's when Charles Berlitz (infamous for his *Bermuda Triangle* book — see my story in *Great Mythconceptions*) and William L. Moore wrote *The Roswell Incident*, the first book about the events of 1947. This was followed by segments about Roswell on a few TV shows, and a few more books.

Interest really increased on 28 August 1995, when Englishman Ray Santelli released a 17-minute silent black-and-white film of an 'autopsy' of one of the Roswell aliens; the film was shown on the Fox Network. Santelli said that he had been making a film of Elvis Presley's 18 months in the US Army and had been searching the Army's archives. Apparently he was approached by the cameraman who shot the footage of the 'alien autopsy'. They teamed up to produce *Alien Autopsy — Fact or Fiction?*.

This film attracted almost as much interest as the film accidentally taken by Abraham Zapruder of the assassination of US President John F. Kennedy. The alien film showed two pathologists in white contamination suits performing the autopsy, a third person taking notes and a fourth person observing from behind a window. A six-fingered, small, naked humanoid lay on the operating table. As corroborating evidence it was claimed that Kodak had analysed the 16-mm film and verified that it had been shot in 1947.

Sceptics — Special Effects

The Truly Dangerous Company (a special effects company for the movie industry) ridiculed the alien shown in the film.

First, the alien's body 'sagged' in the wrong direction — towards its feet, not the table it was lying on. This is what you would expect if the alien corpse was made from a body-cast done on a living person while they were standing, not lying down.

Second, the arms and legs were never bent past the angle at which the synthetic materials used by special effects people would usually buckle.

Third, the alien's buttocks stayed firm and held its legs and lower back off the table. But a dead person or animal loses all muscle tone immediately after death and sags onto the underlying surface.

Fourth, there were problems with the scalpel incisions. Each time the scalpel cut the skin on the upper body, a dark liquid oozed out of the incision. But once the heart stops, there is no pressure in the circulatory system: dead people or animals don't bleed. (It's easy to fake bleeding, though, by carrying 'blood' to the cut via a thin pipe glued to the side of the scalpel away from the camera.) And when the skin was cut by the scalpel, the skin didn't spontaneously pull apart, leaving a gap — unlike the skin of dead animals or humans.

Other special effects artists made similar criticisms. These individuals include Gordon Smith (special effects in *JFK* and *Natural Born Killers*), Steve Johnson (movie-effects designer from *The Abyss, Species* and *Roswell*) and Stan Winston (special effects on *Aliens* and *Jurassic Park*). So the special effects experts were sceptical.

Sceptics — Forensic Pathologists

Medical pathologists also had a few problems with the autopsy techniques shown in the film.

First, the white-suited characters held their scissors in a tailor grip, which puts the thumb and index finger in the scissor holes. This is quite different from the standard surgical grip, which has the thumb and *middle* (or ring) finger in the scissor holes, with the first finger near the pivot point of the scissors, steadying them.

Second, these so-called 'pathologists' didn't start the autopsy the usual way, which is to move the arms and legs (how else will you find a broken limb?).

Third, they cut very slowly, tentatively and indecisively — which is totally different from the quick, decisive cuts made by pathologists. In every autopsy, the pathologist will always be able to see the tip of the scalpel or scissors, so that they know what they are doing at all times. In this so-called autopsy, the 'pathologists' made many blind cuts. At no stage did they try to follow one plane of one tissue through to its natural end — they just cut straight through everything.

Fourth, the hooded anti-contamination suits worn by the 'pathologists' didn't have any air pipes coming in or out, so their face masks would have fogged over within a few minutes.

Finally, the four individuals present showed very little interest in the internal organs of the alien's gut. This is very unusual for what should be the most important autopsy in the history of the human race. They just quickly and indifferently pulled out the organs (which, strangely, didn't seem to be attached to anything) and plunked them into a bucket.

For comparison, consider the coelocanth. It was an ancient fish that was thought to have died out with the dinosaurs, 65 million years ago. But a live coelocanth was discovered in 1938, near Madagascar. Ichthyiologists (fish scientists) travelled from all over the world to be present at its long, slow and careful dissection, preservation and study. But the supposed alien autopsy was carried out by a few bumbling amateurs in under three hours!

Many pathologists, including Dr Cyril Wecht, ex-president of the US National Association of Forensic Pathologists, are thus very sceptical about the Roswell autopsy.

Sceptics — Camera and Film

If the intention was to record the first autopsy of an alien, why use black-and-white film when colour film was available? Why not record sound, which had been possible for years? And why document this event only with a movie camera, when a still camera with colour film would have gathered so much more information?

Also, the camera technique seems unusually modern and jerky for post-World War II camera operators, who prided themselves on their smoothness. Unusually, the camera was not mounted on a tripod but was hand-held instead. Also very unusual is how the camera went out of focus and became very shaky in the close-ups (where a lot of extra detail would be seen).

So what about the film that was submitted to Kodak? Yes, a film was submitted to Kodak, and Kodak did date it to 1947. But the film that Kodak was given showed images of a staircase and a doorway, as well as some end pieces of film containing no images. In other words, Kodak was never given any images of the autopsy to test. Peter Milson, Kodak's manager for Marketing Planning and for Motion Picture and Television Imaging, said, '… what he's done, obviously I can't blame him for this, is given me a bit of the leader … and said this is the same as the neg, this is from the same bit of film.'

And, finally, how did the cameraman manage to steal the only footage of the autopsy of an alien without the military noticing?

The Autopsy

The word 'autopsy' comes from the Greek words '*auto*' and '*opsis*', and it literally means 'to see for oneself'. This 'alien autopsy' showed us nothing — including the cause of death.

USAF?

Some of the early conspiracy theorists claimed that the Roswell Incident happened at the United States Air Force Base in Roswell, New Mexico in July 1947. But this was impossible.

The United States Air Force (USAF) only came into existence a few months later, on 18 September 1947. It had begun as the United States Army Air Corps on 2 July 1926, and became the United States Army Air Force on 20 June 1941. But at the time of the Roswell Incident, the United States Air Force did not exist.

Originally, some of the early conspiracy theorists actually had 'genuine' documents related to the Roswell Incident, with 'genuine' United States Air Force letterhead. Once the nonexistence of the USAF at that time was pointed out to them, they quietly withdrew their 'genuine' documents.

References

'Special Report. A Surgeon's View: Alien Autopsy's Overwhelming Lack of Credibility', *Skeptical Inquirer*, January/February 1996.

Broad, William, J., 'A suggestion that Dead "Aliens" Were Test Dummies', *New York Times*, 24 June 1997, 1.

Corliss, Richard, 'Show Business: Autopsy or Fraud-Topsy', *Time*, 27 November 1995.

Was the Electric Chair Really Painless?

Every society has ways to keep its members in line. These include silent disapproval, incarceration and expulsion, right up to the death penalty. By the late 1800s, many people were opposed to the cruelty of death by hanging. Usually hanging would kill the victim almost instantaneously — but occasionally it would either slowly strangle, or suddenly decapitate, the prisoner. So when the electric chair was offered as a painless form of execution, many people welcomed it. But, in reality, the electric chair was anything but painless.

New York, New York

In 1834, the New York Legislature almost banned capital punishment because of public revulsion over botched hangings. In a hanging, the prisoner stands on a trap door with a heavy weight tied to their feet; the rope around their neck has about 2 m of slack. The trap door opens, they fall down and, after a couple of metres, come to a sudden halt. The knot of the noose should be tied on one side of their neck, and the combined mass of their body and the weight hanging underneath sends a huge force through the knot. This force tilts their head to the side, away from the knot. If the tilting happens quickly enough, this force snaps some of the vertebral bones in the neck, which cut through the spinal cord and the blood vessels supplying the brain, killing the person almost instantly.

In a few cases, however, the executioner did not tie enough weight to the prisoner's feet, or the prisoner had very strong neck muscles. In these cases, the vertebral bones in the neck did not break and the victim took up to 35 minutes to slowly strangle to death. Sometimes, the executioner tied on too much weight — and the prisoner's head was torn off.

The governor of the state of New York at that time worked out that there were about 35 'practical' ways of executing people, including stoning, crucifixion and lethal injection. He presented these various methods to judges and other influential people for their opinions. Most of them were happy with hanging — but if they didn't like hanging, they preferred electrocution. In 1888, the New York 'Death Commission' recommended executions by electrocution.

War of the Electric Currents

The introduction of the electric chair was accompanied by a secret, and very high-level, business battle. At that time, there were two types of electricity available. If one type was to be associated with death by electrocution, it was believed, people would not want it in their homes.

Some people were already scared of electricity. For example, Caroline Lavinia Harrison was the wife of President Benjamin Harrison, who had the White House wired for electricity. She was so terrified of electricity that she refused to switch off the lights in the White House, or to let her husband do so. So the electric lights would usually burn through the night, even in their bedroom.

Thomas Edison did not invent the electric generator, the electric motor, the electric light or the electrical distribution system that carried the electricity from the generator to the house or factory. But he *was* the first person to combine them into an integrated system — he started the electrical industry. He was a first-rate engineer — and at the time held the record for the greatest number of patents issued to a single person. He also had an enormous ego.

Edison had begun with direct current (DC) electricity, and he stayed with it. His DC network ran at a safe low voltage, but, as a result, lost voltage rapidly over distance. So his network needed a DC generator every kilometre or so.

Edison's rival, George Westinghouse, offered a new product to the consumers: alternating current (AC) electricity. AC could be sent long distances without weakening. Westinghouse was very clever, but he was not in Edison's class, either as an engineer or an inventor. However, he was able to recognise good inventions by other people, and then develop and market them.

Edison resented Westinghouse's success, so not only was there a business war between them, there was also a War of the Ego. Accordingly, Edison secretly did everything he could to give a bad reputation to the electric chair powered by AC. (See 'DC and AC 101' on page 42 for more information about these currents.)

History of Electrocution

Back in 1773, Benjamin Franklin wrote that he had successfully electrocuted a 10-pound turkey, a lamb and several chickens. He did this with electricity that he had stored in six Leyden Jars (an early form of battery). But it took several inventors, all working around the same time, to come up with a working electric chair to electrocute wrongdoers.

One person was the dentist Alfred Southwick of Buffalo, New York, in 1881. He had seen a drunk stumble onto a bare electrical

If you spark me up ...

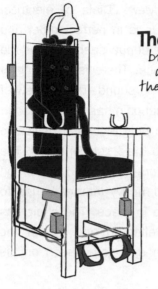

The original electric chair was strongly built from heavy oak, with broad arms and a sloping back. Heavy leather bands held the prisoner's arms, legs and body in place. And the chair was bolted to the floor. The electricity entered and exited the body via wet natural sea sponges on the prisoner's shaved skull and back.

All in all, a very unpleasant way to die.

wire and die immediately. Southwick guessed, with absolutely no evidence at all, that the poor fellow had died painlessly. Southwick was an engineer as well as a dentist, and had served on the commission that had investigated and then authorised the first use of the electric chair.

Two other inventors were Carlos McDonald and A.P. Rockwell. They served on a different New York commission, which was charged with drawing up plans for the first electric chair.

The last person was the inventor Harold Brown, who helped build the first electric chair. But before this he had done much research, all secretly funded by Edison. Brown invited newspaper reporters, engineers and electricians to his demonstrations at the Columbia College of Mines (which later became Columbia University in New York City). He electrocuted some 50 stray cats and dogs, 'proving' that AC was good only for 'the state prison, or the dog pound'.

The Electric Chair Arrives

Eventually, Harold Brown had conducted enough experiments to be quietly confident that his electric chair would work. In 1889, he wrote in the journal *North American Review*: 'Dials of electrical instruments indicate that all the apparatus is in perfect order and record the pressure at every moment. The deputy-sheriff throws the switch. Respiration and heart activity cease. There is stiffening of the muscles but there is no struggle and no sound. The majesty of the law has been vindicated, but no physical pain has been caused'.

Brown's electric chair is probably the only electrical device to stay basically the same from the 19th to the 21st centuries. The original electric chair was strongly built from heavy oak, with broad arms and a sloping back, with some 11 heavy leather bands to hold the prisoner's arms, legs and torso in place. The chair was bolted to the floor. The electricity travelled in and out of the prisoner's body through wet natural sea sponges placed on his shaved skull and back.

Westinghouse knew that Brown was trying to discredit him, and AC, so he wouldn't sell Brown any AC alternators. Instead, Brown got three alternators from dealers in Rio de Janeiro and Boston, helped by Edison's secret funding. In retaliation, Westinghouse secretly paid out hundreds of thousands of dollars in lawyers' fees to try to stop the electric chair execution of the axe murderer William Kemmler. He was unsuccessful, and Kemmler became the first person to be executed in the electric chair in Auburn Prison, New York, on 6 August 1890.

First Electric Execution

Kemmler's death was barbarous. Slightly more that 1000 volts of AC were passed through his body, and he strained against the straps in agony — but he survived. This was because after 17 seconds of watching him, Dr Spitzka, one of the attending doctors, said, 'He is dead', and the electricity was switched off. To their horror, the two dozen people in attendance could see that Kemmler was breathing, groaning, grunting, and still convulsing gently. So the prison officers gave him a second dose of 2000 volts until his body began to burn, which took about 70 seconds. The first unsuccessful dose had dried out some of the electrodes where they touched his skin, so the second dose filled the room with the scent of burning flesh and hair. His blood vessels burst open, and blood squirted out of his skin.

Worse was to come. It took about eight minutes for Kemmler to stop jerking around, but then officials had to wait for his body temperature to drop by the legislated three degrees below normal body temperature for him to be pronounced legally dead. The long dose of electricity had heated him so much that they had to wait for another three hours for him to cool down enough. The autopsy showed that his 'muscles were carbonised, soft tissue resembled well-cooked beef'.

Bad Publicity

At other electric chair executions, the prisoners' leg muscles went into spasms so powerful that they broke the leather straps holding them down. In other cases, the flesh was cooked off the bones.

The problem lies in what the electricity does: it sends the muscles into uncontrollable, and very painful, spasms. It also sends the heart into fibrillation, where the individual heart muscles write in an uncoordinated fashion. But the next burst of electricity would jolt the heart back into a synchronised rhythm, accompanied by massive muscle pain. And back and forth it would go, until the prisoner eventually died.

Edison used his influence to have a newspaper headline written as KEMMLER WESTINGHOUSED. Questions such as 'Do you want electrocutioner's current in your children's bedroom wall?' were bandied around.

Edison had won this war. But, over the years, he lost the battle as AC became the common form of electricity sold in the USA, and the world, because of its technical advantages.

The End of the Electric Chair

The main reason for the abandonment of the electric chair was that it is so painful and barbarous that even hardened Deep South hanging judges were refusing to administer it.

The state of Pennsylvania first used the electric chair in 1915. At its peak usage, about half of the states of the USA employed the electric chair to administer the death penalty. It was also used in the former Republic of China, and in the Philippines. These days, the preferred method of capital punishment is lethal injection.

Over the last century or so, about 4300 Americans have been killed in the electric chair. These days, the proponents of capital punishment assure us that lethal injection is painless. You have to wonder what we'll be told in a few decades from now.

DC and AC 101

Direct current (DC) is what runs your car, body and subway. In DC, the level of voltage always stays the same. The inside of the cells in your body are about 80 mV (thousandths of a volt) positive compared to the outside. Your car runs on 12 volts, while trucks run on 24 volts. Many electric subway trains get the power to run their electric motors from 800 volts DC fed from the steel rails or from overhead wires.

The level of alternating current (AC) is not constant. In one complete cycle the voltage goes from zero to fully positive to zero, down to fully negative, and finishes back on zero again. In Australia, AC has 50 of these complete cycles in each second. But electricity runs on 60 cycles in the USA, and about 400 cycles in passenger jets.

DC vs AC

DC has the advantage that it is easy to generate. However, the generator has brushes that touch a moving part and wear out fairly quickly, which is a disadvantage. An advantage of DC is that it is easy to change the level — you just run it through a resistor. This makes it easy to change the speed of a motor, but the disadvantage is that you burn up energy, which appears as heat in the resistor.

It took the genius of Nikolai Tesla to work out how to generate AC. AC 'generators' are called alternators, because they make alternating current. Alternators do not have brushes to wear out. In the old days (say, pre-1980) it was difficult to change the speed of an electric motor. But modern electronics make speed-changing easy — and you don't waste energy in heating up a resistor.

DC vs AC Transmission

Sending (or transmitting) DC is what killed it as a commercial entity. The two difficulties were that energy was lost in transmission, and that it was messy to change the voltage.

If you want to send a lot of power in a wire, it makes sense to send it as 'high' volts and 'low' current. Low current means that the wire doesn't get too hot. If you want to send lots of current, you have to use a thicker wire, which is heavier and more expensive.

If you send lots of volts (say, 100 000 V) as DC down a wire, the high voltage 'jumps' off the wire and into the air immediately next to the wire. The air molecules get ionised and drift away, and this counts as an energy loss. In a wind, you lose even more energy.

One huge advantage of AC is that it is really easy to bump up the voltage when you want to send it long distances, and bump it down for the home or factory by using a transformer. You can't easily change the voltage of DC.

But sending AC at high voltage down a wire is different. Sure, for a brief instant the wire has created a whole bunch of, say, negatively ionised molecules, but before they get a chance to drift away, the electricity changes its polarity and 'sucks' the negatively ionised molecules back to the wire.

DC is fine for a tiny network, about half a kilometre across. But AC is the way to go for electrical power grids that span a continent. Edison had backed a loser.

References

Jones, Jill, 'New York Unplugged, 1889', *The New York Times*, 13 August 2004.

Moran, Richard, *Executioner's Current: Thomas Edison, George Washington and the Invention of the Electric Chair*, Alfred Knopf, 2002, ISBN 0375410597.

Panati, Charles, *Panati's Extraordinary Endings of Practically Everything and Everybody*, Harper & Row, New York, 1989, ISBN 0060962798, pp 109–112, 156–161.

Can You Restart a Flatlined Heart?

We all know that TV is God. So it follows that medicos on TV must be right up there in the Heavenly Choir. So what happens on TV when the patient's heart shows a flatline (accompanied by a loud, continuous beep), instead of the nice, regular up-and-down heart pattern? The medical team pulls out the defibrillator paddles and tries to jump-start the heart by shoving some electricity into the chest. They succeed, and the next scene usually shows the patient relaxed and happy and thanking the team profusely for saving their life. Totally wrong.

Heart Mechanics 101

The human heart may be the Organ of Love, but it's a darn fine pump as well. Over a lifetime it will pump about 200 000 tonnes of blood.

The heart has four pumps in a series, one after the other. Blood comes into the heart at low pressure and blueish in colour (because it's low in oxygen). The pumps are the right atrium (low pressure, blueish blood), right ventricle (medium pressure, blueish blood), left atrium (low pressure again because the blood has lost pressure going through the lungs, and red because it's now rich in oxygen) and finally the left ventricle (high pressure, red). This high-pressure oxygenated blood is just perfect for supplying your organs.

Each of these pumps is just a set of muscles surrounding a hollow chamber. It has to fill with blood and, within a second, pump that blood out. The timing of these actions is absolutely critical — after all, one pump should push blood into the next pump only when that next pump is empty and ready to be filled.

However, the mechanics of pumping are also incredibly complicated. Think about the individual muscles that wrap around the hollow chamber: their thousands of individual fibres have to contract in the right order, and with the right timing. The end result should be that each contraction of the left ventricle (roughly once every second) squirts out about 80 ml of blood into the general circulation.

Heart Electrics 101

The timing of all the heart's mechanical activity is done by electricity. When your heart beats, a pulse of blood pushes out through all your arteries, making them expand. Once the pulse of blood has gone past, the walls of the arteries shrink again. One very important sensor for controlling the beating of your heart is wrapped around the carotid artery in your neck. It fires off when it senses that the blood pressure is low (when the artery is shrinking) and sends an electrical signal via nerves to a relay station on your heart called the sino-atrial node. The sino-atrial node then passes on the signal until it reaches your heart muscles.

These electrical signals do two jobs: first, they tell each of the four pumps exactly when to fire; second, they coordinate exactly the contractions of the thousands of muscle fibres in each pump.

Finally, as the muscles in the heart contract, they give off electricity. There's a lot of electricity floating around the heart, and you can see these electrical signals associated with the heart muscles contracting on a monitor. The trace is called the ECG (Electro CardioGraph) in Australia and the United Kingdom, or EKG in the USA.

If all goes well, your heart should beat steadily for over 70 years.

Heart Electrics Problems

There are many different problems that you can have with the electrical operations of your heart. After all, it's a long and complicated path from your neck to your heart.

Sometimes the electrical signals stop entirely; this is called 'asystole'. The ventricles stop contracting, and (of course) the heart stops pumping blood. The ECG shows a flatline. This is the close-up image you usually see on TV.

Sometimes the electrical signals become uncoordinated. This is called 'ventricular fibrillation' (VF). In VF, the individual tiny muscles in the ventricles keep beating, but they are not coordinated. If you cut open the chest in such a case and gaze at the heart, it looks just like a bag full of wriggling worms. When one tiny muscle contracts, the one next to it might relax, so the blood just shifts around inside the ventricle and doesn't leave the heart, a bit like a car with a powerful engine spinning its wheels. Lots of energy is being used up, but it's all wasted internally. The ECG now shows a very chaotic rhythm, and again the heart stops pumping blood.

Sometimes the heart beats very quickly. This is called 'ventricular tachycardia' (VT). Again, the heart doesn't pump any blood into the arteries, because the blood doesn't have time to get into the ventricles — the filling time is too short. In VT, the ECG shows a very fast rhythm.

VF and VT are grouped together, because the treatments are very similar.

Heart Mechanics 101

The heart has 4 pumps that work one after each other

Blood comes into the
heart 'blue' in colour
and at low pressure

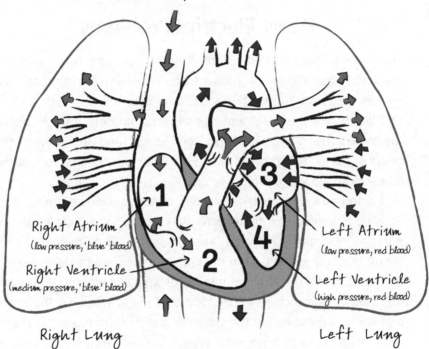

Right Atrium
(low pressure, 'blue' blood)

Right Ventricle
(medium pressure, 'blue' blood)

Right Lung

Left Atrium
(low pressure, red blood)

Left Ventricle
(high pressure, red blood)

Left Lung

The Organ of Love

(and its 'hood ... not drawn to scale)

Each pump is nothing more than muscles surrounding
a hollow chamber. Each chamber has to fill with blood
and pump it out within about a second.
All going well, with each contraction of the left
ventricle, 80 mls of good red blood enter the
general circulation.

Defibrillator

Now, here's the important part. In both VF/VT and asystole, the heart stops pumping blood. But the treatments are very different.

In VF/VT, the treatment is the famous jumper cables and paddles — the defibrillator (which some ambos call a 'Packer Whacker', after businessman Kerry Packer donated defibrillators to the NSW Ambulance Service). The paddles are really just electrodes; they are placed so that the heart is roughly between them. When you hit the Go button, the paddles deliver about 300 joules of electrical energy in a very short period of time — about 4–12 milliseconds. The electricity passes in from one electrode through the skin to the heart and out through the skin to the other electrode. The electrical shock delivered to the chest (and the heart inside) is similar to gently slapping a hysterical person on the face to bring them back to their senses. The electrical shock stops the rapid chaotic electrical activity — the VF/VT. Then, hopefully, the sino-atrial node in the heart restarts in its regular rhythm, and 80 ml of blood squirt out every second.

Defibrillation works *only* if there is already electrical activity going on in the heart.

In asystole, the ECG shows a flatline and the standard treatments are not very effective. They involve the full gamut of life support, including CPR (cardiopulmonary resuscitation) and drugs. Blasting the heart with electricity is one treatment that definitely does not work.

One 1990 study in Sydney analysed some 1339 cardiac arrests (heart attacks). In about 40% of the cases, the patients had VF or VT. In about 30% of cases, they had asystole. Only 97 left the hospital alive — and of these, all except three had VF.

So the treatment for asystole (flatline ECG) is not the defibrillator — and no matter what they say on TV, if it's asystole, it ain't good.

Resuscitation
by Breathing

There's a very long history of bringing people back to life by helping them with their breathing.

Four thousand years ago, Egyptian mythology described how the goddess Isis brought Osiris, her husband, back to life by breathing into his mouth. Back then, the Egyptians would get direct access to the windpipe (trachea) in a breathing emergency by cutting a hole in the neck and blowing air through a reed. In 356 BC, the writer Homer described how the Greeks did breathing resuscitation — by opening the trachea (windpipe) to relieve choking.

Resuscitation and
Heart Problems

When you're resuscitating people, heart problems are harder than breathing problems to understand and deal with. It took until the first century AD for Pliny the Elder to be the first to describe his fellow citizens dying suddenly from heart problems.

The next big jump in knowledge happened in 1775, when Abildgaard (a Dutch veterinarian/physician) experimented with chickens. He showed that, by using electric shocks from a Leyden Jar, he could first stun the chickens, and then revive them. In 1850, Drs Hoffa and Ludwig showed that they could excite the heart of a mammal into VF, and the animal would die.

In 1911, Dr Hoffman recorded the first ECG of a human as he suffered VF. In 1939, both Drs Hamilton and Robertson and, separately, Drs Smith and Miller actually recorded VF in patients who died from VF. In 1947, Beck corrected VF in a 14-year-old boy with alternating current (AC) directly to the heart muscle.

In 1956, Zoll was able to correct VF with direct current (DC) electricity to the chest wall. The first advantage was that DC was much easier to deal with than AC. Second, it was much better now that they did not have to cut open the chest wall.

Defibrillation for VF had finally arrived.

Reference

Adgey, A.A. Jennifer, 'Resuscitation in the past, the present and the future', *The Ulster Medical Journal*, May 2002, pp 1–19,

Carrots and Night Vision

Parents have lots of tricks to get their kids to eat their vegies. One often-used persuasive tale is that if the kiddies don't eat their carrots, they'll go blind or, at the very least, not be able to see in the dark. The parents are telling a little white lie. But there is in this little fib, as there is in all good mythconceptions, a small seed of truth.

History of Carrots

The carrot was originally bred from a common weed, Queen Anne's Lace, somewhere in the Middle East, probably near Afghanistan. Carrots had reached the Greeks and the Romans before the birth of Christ. The Greeks called the carrot *philtron*, and thought it would make men and women more interested in loving-each-other-very-much-in-a-very-special-way. Caligula, the mad, bad and dangerous Roman Emperor, believed these stories and made the men of the Senate eat carrots because he wanted to see them 'rut like wild beasts'.

By the 13th century, carrots had reached northwestern Europe, India, Japan and China. Back then, carrots came in many colours

— red, black, purple, white and yellow. The Dutch then bred them to produce the bright orange carrot we know today.

During the Middle Ages in Europe, carrots were used as a medicine to cure most maladies, including sexually transmitted diseases and snakebites. In Elizabethan England they were used both as food and finery — some citizens decorated their hair, hats and clothing with the stalks.

Carrots and Vision

Parents are partly correct — carrots are involved in vision, thanks to their link to vitamin A. Vitamin A is one of the fat-soluble vitamins and it travels to the eye in the blood. The eye is where incoming light is turned into electricity. In the eye, vitamin A is slightly changed to make a chemical called retinal. When light hits retinal, it gives off electricity, which ultimately ends up in the visual centres in the brain, which mysteriously turn electricity into vision.

So no vitamin A means no retinal, which means no electricity — and no vision.

You can get vitamin A from some animal products (such as fish and liver), but not from the plant kingdom. However, yellow/orange vegetables and fruit have a coloured chemical that can be turned into vitamin A. These chemicals are called 'carotenes', and they cause the yellow/orange colour of some fruit (apricot) and vegetables (pumpkin, sweet potato), as well as the yellow/orange of the feathers of canaries and the shells of lobsters. Your liver converts carotenes into vitamin A. (By the way, it's actually more complicated than that, which is just what you'd expect from the human body. There's not just one single vitamin A but, rather, a family of very similar chemicals that all go under the name of vitamin A. For example, vitamin A1 has two more hydrogen atoms than vitamin A2. But they all work pretty much the same.)

If you don't have enough vitamin A, you won't have healthy skin, a healthy immune system — and you won't have good vision.

Carrots and World War II

So if you don't get enough carotenes or vitamin A in your diet, eventually you will suffer problems in your vision. This was the basis of the myth started by the British Royal Air Force (RAF).

In the Battle of Britain in 1940, the British fighter pilot John Cunningham became the first person to shoot down an enemy plane with the help of radar. In fact, in World War II he was the RAF's top-scoring night fighter pilot, with a total of 20 kills. (After World War II, in 1948, he set a new world altitude record in a Vampire fighter-bomber, and in July 1949 he made the first test flight in the first passenger jet, the Comet.)

Some pilots, like Jonnie Johnson and 'Sailor' Malan, were better flying in daylight. But others, like Group Captain John Cunningham, were better at night. He was very good at thinking in three dimensions and his nickname (which he never liked) was 'Cats' Eyes'.

Wassup doc?

What they said about me was all lies. How can I move on?

The carrot, a good source of vitamin A, but not as good as some have said at increasing your night vision.

The RAF in WWII invented a myth that carrots (thanks to vitamin A) were the reason for their pilots' exceptional night vision. This was to lead the enemy away from the real reason ... that they had radar!

The RAF put out the story in the British newspapers that Cunningham and his fellow night pilots owed their exceptional night vision to carrots. People believed this to the extent that they started growing and eating more carrots, so that they could better navigate at night during the blackouts that were compulsory during World War II.

It Ain't True ...

It is true that not enough vitamin A will make you unable to see at night. But, conversely, lots of vitamin A can be poisonous — and it definitely will not give you super-human vision at night. The whole story about the pilots and their carrots was a myth invented by the RAF to hide their use of radar (from **RA**dio **D**etection **A**nd **R**anging). It was radar that really located the Luftwaffe bombers at night — not human carrot-assisted super-vision.

In a radar unit, an antenna sends out a brief burst of radio waves (with a peak power, say, of one million watts), then stops transmitting. It then listens for a very weak echo as some of the radio waves bounce off the metal skin of the plane and return to the antenna. These returning radio echoes can be as weak as one-trillionth of a watt. The time interval between the transmitted and received signals gives the distance to the plane.

At the beginning of World War II, France, the UK, the USA, the Soviet Union, Italy, Japan and Germany had each explored and researched radar, but only the UK had developed a fully functioning network. This network was called 'Chain Home' and operated 24 hours a day from September 1938 until the end of the war.

Why, though, did the German Air Force see past the obvious radar towers on the English coast and fall for this blatant 'carrot super-vision' myth? Because this myth that carrots would improve eyesight already existed in German folklore.

Light into Electricity
(Complicated Version)

If you don't have vitamin A, you don't have vision.

Once you have some vitamin A in your body, it travels in the blood to the retina, which is a thin sheet (about 0.3 mm thick) on the inside of each eye. The retina is where light is turned into electricity. The retina has one layer of cells to generate this electricity, and about nine more layers of cells to process it. (So the retina is not just a dumb converter of light to electricity, but instead an intelligent processor.)

The retina's first layer has about 130 million cells called rods, which deal with low-level night vision, and which are called rods because they look like little rods. It also has about seven million cells called cones, which deal with high-level daytime vision, and which are called cones because they look like little cones. The ends of rods and cones are full of closely packed flat membranes which are covered with a chemical called rhodopsin.

Rhodopsin is made up of two parts: one part is a protein that is not sensitive to light; the other part is a chemical called 11-cis-retinal. When a single photon of light hits this second part, it changes shape into a chemical called 11-trans-retinal, and this change activates a chemical called transducin. The presence of transducin inside the rod or cone then sets off a series of biochemical reactions, which finish with the release of some neurotransmitters from the end of the rod or cone. The neurotransmitters change the electrical activity of the next layer of cells in the retina. This layer of cells then 'talks' to the next layer, and so on — all the time doing some very fancy processing and compressing of the data.

After the electricity is fully processed in the nerves of the retina, it then travels along the two optic nerves (one from each eye) to the visual centres at the back of the brain. These visual centres turn the electricity into the strange wall-to-wall sensation that we call vision.

Why Night Vision is Actually Blue and White

There are three types of cones: some are sensitive to red light, some to green light and some to blue light. But daytime vision has not just three colours but millions of colours. It does this by combining the signals from the red-sensitive, green-sensitive and blue-sensitive cones. The seven million cones can function only when there is lots of light around, such as in daytime. At night there is not enough light for them to work, so the rods take over.

The 130 million rods cannot work in daytime because the high levels of light 'saturate' them. But they work just fine at night. It takes them about 45 minutes to reach their full sensitivity. There is only one type of rod, which is sensitive only to blue light. This is why night vision is just different levels of brightness of a bluish colour. Dark blue is close to black, which is why we usually think of night vision as black and white.

By the way, you can actually see this happen if you sit outside at sunset, away from artificial lights. As it gets darker, your cones will stop working, and the rods will gradually kick in. So you will see all the colours gradually fading into blue.

Ancient Cancer

If you read some of the New Age magazines, you will be astonished to discover that cancer is a brand-new disease. These magazines claim that cancer is a disease created entirely by our nasty environment, which is totally loaded with carcinogens (including chemical toxins, pollutants and poisons) and deadly radiation and magnetic fields. As part of trying to sell you their preventer/curer of cancer, they will argue that in the good old days, before we humans developed industrial societies, there was no such disease as cancer.

Cancer in the Past

As 'evidence' the mags will quote a *National Geographic* study that examined the bones of 3160 Croatians who died between 5300 BC and the mid-19th century. In the first paragraph, John Pickrell wrote that '… Croatian archaeological collections suggest that cancer is more common today than at any point in humankind's history.'

However, we definitely know that cancer existed in the distant past. Dr B.M. Rothschild (an American radiologist) X-rayed over 10 000 dinosaur bones (averaging 70 million years old) and found clear evidence of tumours. In one group of 97 hadrosaurs (duck-billed dinosaurs), his team found 29 tumours. Rothschild thinks that this is because hadrosaurs ate conifers, which are rich in carcinogenic chemicals. (Nasty chemicals have been around for a long time, as have nice chemicals.) Indeed, cancer is so common to living creatures that it has been found in budgerigars and coral.

Cancer 101

The simple word 'cancer' covers a wide group of several hundred diseases, which all have one common factor: the cells grow without limit. In your body, normal cells grow to a certain stage and then usually divide to replace old cells in a controlled fashion, often because of signals from local normal cells. But abnormal cancerous cells will keep on growing and growing without limit. Cancer cells can cause damage by invading normal structures, or by releasing chemicals into the body, or by dumping more cancer cells into the bloodstream to seed and grow elsewhere in the body.

All cells (apart from red blood cells) have DNA (deoxyribonucleic acid), which acts as a blueprint for making another cell. Cancers happen when there is an abnormality in the DNA.

There are many known causes of damage to the DNA. First, when a cell divides it has to make two copies of the original DNA, which has some three billion separate elements (like rungs in a ladder). Mistakes can happen in this copying process. Second, there are environmental agents that can damage DNA. They include radiation (sunlight, X-rays, etc.), chemicals (such as in conifers), viruses (such as the hepatitis B virus, which can cause hepatocellular cancer), and even by-products of your own internal metabolism (such as free radicals). Third, damaged DNA can be inherited (e.g. hereditary retinoblastoma).

Regardless of how it began, there is now an abnormality in the DNA in one single cell. This cell then transmits this abnormality to its two 'daughter' cells when it divides. Of course, there are many levels of control (checks and balances, fail-safe and self-destruction mechanisms) to stop the cell with the abnormal DNA from dividing. But every now and then, one such cell with abnormal DNA sneaks through. After some 30 such divisions, there is now a

Cancer - the basics

The word 'cancer' covers a wide group of hundreds of diseases. All have ONE common factor – the cells grow without limit. In your body, normal cells grow to a certain stage and then divide to replace old cells in a controlled fashion.

but ...

Abnormal cancerous cells will keep on growing without limit. Cancer cells cause damage by invading normal structures, or by releasing chemicals into the body, or by dumping more cancer cells into the bloodstream.

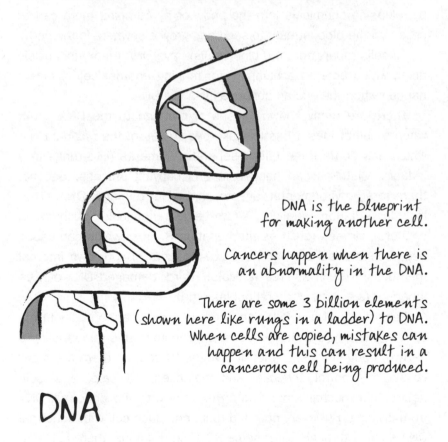

DNA is the blueprint for making another cell.

Cancers happen when there is an abnormality in the DNA.

There are some 3 billion elements (shown here like rungs in a ladder) to DNA. When cells are copied, mistakes can happen and this can result in a cancerous cell being produced.

DNA

lump of some billion cells, weighing about one gram — and at around this size, the new cancer is clinically detectable.

History of Cancer

The New Age magazines are very wrong — we humans have known about cancer for a long time.

Five-thousand-year-old mummies from Peru and Egypt show boney cancers and about 2400 years ago the Greek doctor Hippocrates, who gave us the Hippocratic Oath, also gave us the name 'cancer', from the Greek word *karcinos*, which means 'crab'. He thought the vice-like grip of the crab's claws resembled the infiltrating filaments that some cancers have — and, of course, the terrible pain. Around 200 AD, the Greco-Roman physician Galen thought that cancer was caused by inflammation.

Hereditary factors were first considered as a possible cause of cancer around 1745. In 1761 the English physician John Hill was the first to point the finger at environmental chemicals, when he found a link between nasal cancer and snuff (a form of tobacco). In 1775 Sir Percival Pott brought occupational health into the picture when he published his well-known finding of the link between cancer of the scrotum and the job of chimney sweep. As chimney sweeps crawled up and down the chimneys to clean them, they drew carcinogenic chimney soot into their trousers.

Today, in the wealthy countries, one person in every three will have a cancer at some stage in their life. In most of these countries, cancer is the leading cause of death. There is a massive surge in cancers in people aged between 55 and 75 — simply because of the extra time that the cancer has had to grow, increased exposure to environmental factors, and so on.

The Myth

Humans' longer lives are the reason why cancers are so much more common today than in the Croatians of 7300 years ago.

Back then, the average life expectancy was only 36 years of age, which meant that people died before the slower-growing cancers had a chance to kill them.

DNA — Other Stuff

DNA stands for **d**eoxyribo**n**ucleic **a**cid.

Physically it looks like an incredibly long, but very skinny, ladder. Like your common household ladder, it is a few metres long. Like your common household ladder, it has two side rails. But while your common household ladder might have ten rungs, DNA has about three billion rungs.

One of the great discoveries of the 20th century was that any three of these rungs can act as a blueprint to tell the organic machinery inside each cell to make an amino acid. There are about 20 or so amino acids. If you join a bunch of amino acids together, you get a protein. Proteins can either stay inside the cell (to make it bigger), or leave the cell to be exported around the body (such as hormones, like insulin or thyroid hormone).

But DNA is not just a blueprint to make another cell or various chemicals. It is also a medical textbook and a history book of our genetic heritage, plus a record of how we have survived various epidemics, as well as …

Cancer — Single Disease?

I always get a little shudder when I hear those four fateful words on the TV evening news and current affairs programs: 'new cure for cancer'. If the reporter is male, they will usually speak in a curiously high-pitched yet nasal voice, expelling only three syllables at a time as they beat one forearm in a metronome-like fashion. It also helps if the reporter has haemorrhoids, to give him an undeniably sincere look of anxious concern. Unfortunately, the 'new cure for cancer' is about as real as many of the other things that you see on TV.

The first problem with the TV reporter's ridiculous claim about a 'cure for cancer' is that cancer is not a single disease. It's actually the name for a whole family of diseases that all share a common characteristic: the uncontrolled growth of abnormal cells.

There are over 100 separate diseases that come under the name of cancer, such as lung cancer, pancreatic cancer, breast cancer, brain cancer, prostate cancer, and so on. And each of these cancers has many sub-types — for example, lung cancers include small cell cancer, squamous cell cancer, large cell cancer and adenocarcinoma.

Saying that we have a cure for cancer is like saying that we have a cure for infectious disease, which as a family includes the deadly Ebola virus, influenza, AIDS and the common cold (rhinovirus).

The second problem with such a bold prediction of a cure is that each of the many different types of cancer has its own separate set of causes, treatments and natural history. Regardless of what the TV reporter says, a 'new cure' for one type of cancer will not cure another cancer. And this leads to the third, and main, problem.

The third problem is that we don't understand enough about cancer to really cure it. Mostly, our treatment of cancer is similar to so-called primitive agriculture: slash and burn. We 'slash' the cancer with a knife to remove as much as possible, and then we 'burn' what is left with drugs or radiation, or both. Each new advance in cancer treatment gradually increases the percentage of people who survive that specific cancer to (say) five years — the so-called 'five-year survival rate'.

So what the TV reporter really means is that the five-year survival rate for just one of the several hundred types of cancer has been increased from 30% to 40%, by tweaking the combination of surgery, drugs and radiation. And occasionally medicine is lucky enough to stumble across a drug or combination of drugs that are very good at killing a specific cancer, for example, certain childhood leukaemias. The reporter definitely does not mean that medicine has now discovered how to cure all types of cancer, with a 100% survival rate.

However, as the Genetic Revolution rolls on and we begin to understand more about our DNA, we will then get closer to real cures for cancer.

Reference

Metastatic cancer in the Jurassic', by Bruce M. Rothschild, Brian J. Witzke, Israel Hershkovitz, *The Lancet*, Vol 354, July 31, 1999, pp 398.

'On the offensive', by Alison Abbott, *Nature*, Vol 416, 4 April 2002, pp 470–474.

'Epidemiologic study of tumours in dinosaurs', by B.M. Rothschild et al, *Naturwissenschaften*, Vol 90, 2003, pp 495–500.

'Ancient Skeleton Collection Yields Cancer Clues', by John Pickrell, *National Geographic News*, 13 July 2004.

Do Cats Really Purr When They're Happy?

Domestic cats have lived with people for many thousands of years. It's a simple and happy relationship — we give them food and shelter, and they catch mice and rats for us. Cats let us know that we have kept our side of the bargain by purring, which we all firmly believe is a sign of contentment. But, once again, it's just not true. Cats also purr when times are bad.

Who Purrs?

The sound of a cat's purr is very distinctive. Cats start purring when they're about a week old, and they can purr both while breathing in *and* breathing out, and with their mouth closed. Baby kittens can even purr with a mouth full of nipple while guzzling milk. (The kittens find their favourite nipple by the smell they left on it last time they had a feed.) The cat mother knows that all her kittens are feeding well by their regular purrs, and they know that everything's fine by her purr. In this case, purring is a feel-good signal.

It seems that cats can either purr or roar — but not both. Cats that purr (e.g. domestic cats, mountain lions, pumas) can't really roar well, and cats that have a terrifying roar (such as tigers and lions) can't really purr. As Charles Darwin wrote in 1872: 'The puma, cheetah and ocelot likewise purr; but the tiger, when pleased, emits a peculiar short snuffle accompanied by closure of the eyelids. It is said that the lion, jaguar and leopard do not purr.'

If you try to copy a cat's purr, you can make a kind of rumble while you breathe out. When you breathe in, the noise you make sounds rather strangled — and nothing like a cat's purr.

Purr is a Mystery

We've been able to get to the moon and back, we have mapped the human DNA, and we also invented income tax. But we still don't fully understand, even in the 21st century, the mechanics of a cat's purr.

Pretty Crummy Theory

One often-quoted theory is that the cat's purr has nothing to do with the cat's larynx (the voice box). Instead, the purr is said to come from the turbulence of the blood going into the heart. Supposedly, the cat can somehow apply a little pressure to the vena cava, the huge vein that feeds blood into the heart, and this constriction then causes turbulence in the blood, which is somehow amplified. It is claimed that various structures around the lungs act as a kind of amplifier, and this sound is then supposed to be further amplified by the sinuses in the cat's head.

There is no scientific proof for this theory.

The purr-puss of purring

purr

It's often thought that a cat's purr is a sign of contentment. BUT cats also purr when times are bad.

In the early '70s, it was discovered that air going in and out of a cat's lungs was being 'jittered' many times per second by some muscles along the air pathway.

The purr-o-meter

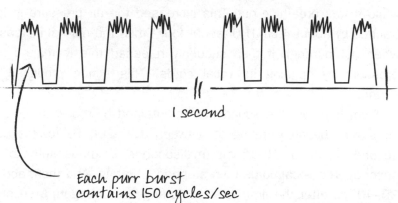

30 purr sound bursts per sec

1 second

Each purr burst contains 150 cycles/sec

Better Theory

Back in the early 1970s, Drs Remmers and Gautier, from the Department of Physiology at Dartmouth Medical School in New Hampshire, studied what they called 'feline purring'. From their reading and experience, they figured that the air going in and out of a cat's lungs was being mechanically 'jittered' many times per second, by some muscles, somewhere along the air pathway.

Considering the fairly primitive technology available back then, they did very well.

They monitored the air pressure in the trachea (windpipe) of their cats, and monitored the actual purring sound with a sensitive microphone and oscilloscope. They also looked for muscular activity. When muscles contract, electrical signals are given off. So they inserted tiny electrodes to pick up electrical signals from the muscles of the larynx, the chest and the diaphragm. The diaphragm is a curved sheet of muscle between your lungs and your gut; when it contracts, it pulls the bottom of your lungs downward. Air then rushes into your lungs via your nose or mouth.

Better Results

Remmers and Gautier found that purring started with the cat almost totally closing the gap between the vocal cords in its larynx while it was breathing out. This increased the air pressure in its trachea by about 4 cm H_2O (about four times the normal increase when not purring). It then suddenly released this increased air pressure by opening its vocal cords. This action produced a sound.

From here on, the sound was maintained by the vocal cords. First, they began vibrating at between 150–200 Hz (cycles per second) for about 10–15 ms (milliseconds, or thousandths of a second). The vocal cords then stopped vibrating. And then, about 30–40 ms after the first burst of purring sound began, a second burst began. The purring could then go on for minutes, or even hours, with about 25 to 33 of these bursts of sound being produced every second. Even though the cat was producing a burst of sound, then a silence, then another burst and so on, our ears would hear this as a continuous purr.

For each of the cats they measured, the length of each tiny burst of sound (around 10–15 ms long) was incredibly regular; the greatest variation was only 1 ms. However, each cat would make its purring louder or softer by increasing or decreasing the intensity of the electrical signals being sent to the muscle.

These electrical signals were coming from a pacemaker somewhere in the brain. Drs Remmers and Gautier were not sure exactly where this pacemaker was.

When the cat was breathing in, both the larynx and the diaphragm would show these regular 150–200 Hz bursts of activity, but when the cat was breathing out, only the larynx would jitter away electrical activity. In this case, the air would be driven out by elastic recoil of the lungs, and the diaphragm would go along passively. If you listen carefully, you can hear a very short interruption to the regular purr as the cat changes from breathing in to breathing out, and vice versa.

If you want to be inelegant, you can think of the cat's purr as a cute version of the human snore, but actively driven by a pacemaker in the brain.

Why Purr?

Cats certainly purr when they are contented. But they can also purr when they are in pain, and giving birth, and being threatened. A cat giving birth may be in ecstatic happiness, so a purr is possibly understandable. But a cat in pain is a cat in pain — and yet it will sometimes purr. Sometimes, cats that are dying will purr. These cats are definitely not contented.

Desmond Morris, a British zoologist, says that purring in a cat can be related to friendship. Either they are content and offering friendship, or they are anxious and are asking for friendship.

Some veterinarians think that purring is a form of meditation — something that you do when you are happy or sad, threatened or content. If they can get the funding, they would love to do an MRI study on a purring cat's brain.

A cat's purr is a bit like a human smile. Sure, we smile when we are happy — but we might smile nervously when we feel threatened, or if somebody is being catty towards us.

We don't really know the purr-pose of the cat's purr …

Purr = Healing Agent?

Some people have made vast leaps of faith about cats' purr, based on our very small amount of knowledge. They claim that a cat's purr can heal.

You can read vague statements such as 'the cat's purr increases the efficiency of the circulatory system, and so keeps the cat healthy'. Veterinarians tell me that there is absolutely no proof for this.

The spiel goes that the purr of the cat heals/realigns/strengthens its bones/organs/mental state. Therefore, some people claim, you should buy their CD of cats purring, which will heal all your ills. This is mixed in with some pseudo-scientific jargon about 'healing frequencies', and then some random numbers are quoted.

The reality is that there is no proof (yet) that the cat's purr does anything curative for the cat. There is also no proof (yet) that the cat's purr does anything for humans.

However, in some cases, powerful vibrations applied directly to the feet of older people (via special shoes) have improved their balance. Does this have anything to do the cat's purr? Nobody has proved anything — yet …

References
Morris, Desmond, *Cat Watching,* Three Rivers Press, 1993, ISBN 009178977X.

Remmers, J.E. and Gautier, H., 'Neural and Mechanical Mechanisms of Feline Purring', *Respiratory Physiology*, 1972, Vol 16, pp 351–361.

Breathing and Oxygen

We all know that breathing is not an optional activity. Every few seconds, for our entire lives, we have to take a breath. Nearly 2400 years ago, the great Greek thinker Aristotle wrote his treatise *On Breath*, in which he asked the very deep question: 'How can we account for the maintenance of breath inherent in us?' It's a good question: why do we breathe?

Most of us think that we breathe because we are low in oxygen. But most of us are wrong, and this may explain why fit young people can fail to come up for air in a back-yard pool — and die.

History of Breathing

Aristotle (384–322 BC) had his own special theory about why we breathe. He reckoned that it was needed to remove the massive body heat manufactured by 'the fiery nature of the soul which exists in the heart'.

Galen (129–216 AD), the Greek physician who lived after Aristotle, realised that there was probably a 'pacemaker' at the top of our spinal cord. He noticed that gladiators (and animals) injured below the top of the neck could still breathe. But if they suffered an injury very high up in the neck, they stopped breathing immediately.

By the 1840s, Dr Flourens had come to think that the 'breathing centre' was in the brainstem — somewhere around the bottom of our skull and the top of our neck. Indeed, he had even narrowed it down to a specific part of the brainstem called the medulla.

Carbon Dioxide — Theory

Modern physiology has taken this knowledge further.

We do have a central 'pacemaker' that drives our breathing: a collection of networks of neurons in our hindbrain. One network fires off when we breathe in, another one fires off when we breathe out, and a third one fires off both when we breathe in and out. The networks all work together to make up our central pacemaker. This central pacemaker reacts to chemical signals from around our body — from chemoreceptors in the arteries, and inside the brain. These chemoreceptors respond both to oxygen and carbon dioxide.

Low oxygen levels will act as a minor trigger to your pacemaker. But, surprisingly, the main trigger to breathe is a high level of carbon dioxide in your blood. This absolutely amazed me when I learnt it in my first-year medical studies.

We are carbon-based creatures and we eat carbon-based foods; our foods have carbon atoms in them. The carbon atoms are strung together, making molecules. We break these carbon molecules apart, extract the energy that holds these molecules together and use that energy to run our daily metabolism. As part of the process, our cells marry a single atom of carbon to two atoms of oxygen to make carbon dioxide, which we breathe out of our mouths as a waste product.

We absolutely have to get rid of this carbon dioxide, so carbon dioxide is the main trigger to keep us breathing.

Carbon Dioxide — Practical

I had learnt this information about carbon dioxide academically in physiology classes, but I really understood it only too well when I nearly killed myself on live television (*Good Morning Australia*, Channel 10, 1990).

I was explaining why helium gives you a squeaky voice and had prepared myself by hyperventilating helium gas for a minute in the ad break, before the camera swung onto me. Hyperventilating is breathing in and out quickly and deeply, and I wanted to make sure that my lungs were absolutely loaded with helium. As we rolled into the program, I began speaking in a squeaky voice, explaining why helium had this effect. After a minute or so, I finished the explanation. The camera (and everybody's attention) swung over to the hosts, Kerri-Anne Kennerley and Tim Webster,

Take a deep breath ...

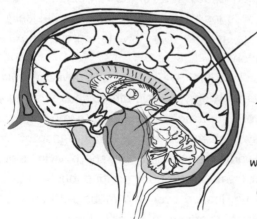

The approximate location of the central 'pacemaker' that drives our breathing.
It's a collection of neurons in our hindbrain. One network fires off when we breathe in, another one fires off when we breathe out, and a third one fires off both when we breathe in and out.

A cross-section through the grey matter (or for those with a little more knowledge ... the 'Midsagittal View').

on the Home Base. Without the camera's glare, nobody was paying any attention at all to me.

I suddenly noticed that I was very slowly crumpling to the floor. I choreographed my slow-motion fall so that I didn't make any noise (this was Live Television, after all, and I was a professional Weather Man with standards to uphold). And I didn't want to hurt myself either. Then, as I lay on the floor, my vision started to fade. This was because the retina is a very oxygen-hungry organ, and my retina was being starved of oxygen. I gradually realised that I was passing out, and probably even dying. A very small part of my logical brain wondered why on earth this should happen.

The Answer

Suddenly, I remembered my first-year respiratory physiology. I realised that I wasn't breathing because when I had vigorously hyperventilated, I had blown all the carbon dioxide out of my lungs and out of my bloodstream. No carbon dioxide meant no desire to breathe; no breathing meant that I was very low in oxygen — and low oxygen meant that I was losing consciousness. So basically I wasn't breathing because I was very low in carbon dioxide (because of hyperventilation).

I tried to breathe, but I couldn't — it felt really awkward, like trying to take another breath when you have already just filled your lungs with air. As the world kept colour-shifting into more grey, I tried again, and suddenly I felt myself taking a huge gasp of air. Almost immediately, the world snapped into full, sharp, vivid colour again, and I realised that I was on the floor, panting loudly. One of the camera crew heard me, looked over and motioned me to be quiet by frowning and putting a very stern finger to his lips.

There have been some recent cases of fit young people dying in back-yard pools, while trying to hold their breath for as long as possible. Perhaps some of them hyperventilated first and blew off all their carbon dioxide — and, along with it, the essential trigger needed for breathing. Then, as they lay on the floor of the pool, without the urge to breathe, they died without taking a single breath. And if they had tried to breathe, they might have gulped in a lungful of water ...

So, be very careful when you have a competition to hold your breath underwater while skylarking in the back-yard pool.

Add Carbon Dioxide to Oxygen?

Oxygen is possibly the most commonly administered 'drug' in the world. But Drs Steve Iscoe and Joseph A. Fisher think that we might need to change how we give it — by adding some carbon dioxide.

The normal air we breathe has about 21% oxygen; medical oxygen can be 100% oxygen. Dr Iscoe explains: 'Pure oxygen can reduce blood flow to organs and tissues by increasing ventilation. The increase in ventilation, which is almost never considered, 'blows off' carbon dioxide, and this fall constricts blood vessels. When carbon dioxide is added, however, the blood vessels dilate, increasing blood flow and causing more oxygen to reach tissues in key areas like the brain and heart.'

Carbon dioxide is our main trigger to breathe. And high levels of carbon dioxide make you breathe faster, in an effort to blow off the excess carbon dioxide.

Dr Iscoe thinks that adding carbon dioxide to oxygen could benefit patients suffering from heart attack, stroke and carbon monoxide poisoning. It could also assist people with Type II diabetes who have foot ulcers and gangrene as well as hospital patients with wounds infected with drug-resistant bacteria, and could perhaps improve cerebral blood flow to a baby during a difficult birth.

Carbon Dioxide in the Blood

The system that carries carbon dioxide in the blood is remarkably complicated.

About 5% of the carbon dioxide is dissolved in the blood as gas. The remaining 95% of the carbon dioxide causes various chemical reactions with the liquid and cells in the blood.

Each time you breathe out, you lose less than 10% of the carbon dioxide in the blood. This means that the acidity of the blood hasn't changed too much between entering and leaving the lungs.

Freediving

There's a dangerous new ocean sport called freediving. People try to dive as deeply as they can, without any breathing apparatus, and they can stay underwater for over 7.5 minutes.

This can be accomplished thanks to a physiological process called the Diving Reflex. When you place your face in cold water there is a massive shift in your blood circulation, to preserve blood for the essential organs. The blood supply is closed down in most of your body, except for your brain and your heart muscles. This reduced oxygen demand means that your available oxygen will last longer. And, like most physical activities, you can train to hold your breath longer for diving.

In 'no limits' freediving, people hang onto a weighted sled as it plunges towards the ocean floor. The current record is over 150 m, and the pressure at that depth is 15 atmospheres (150 tonnes per square metre). Your lungs shrink from 6 litres to the size of a soft-drink can. The pressure squashes the spleen and forces red blood cells into the circulation. And, of course, you can't see in the absolute blackness, the temperature is 4°C, your veins have collapsed, and your ears and sinus cavities are in terrible pain. Freedivers can go from the surface to the deep and be back on the surface within seven minutes. But traditional divers who dive to 150 m have to spend five days decompressing.

There have been many deaths in freediving. Many of them happen as the divers come up through the last 10 m to the surface. As the lungs double in size in that last 10 m, the oxygen levels in the blood drop and the divers can suffer 'shallow water blackout' as they faint.

Reptile Record

The deepest dive for a reptile on record is 640 m. The reptile was a leatherback turtle, which can grow to 2 m long, and can weigh up to 916 kg. Their main food is jellyfish and gelatinous zooplankton. Normally they would not dive deeper than 200 m, because it takes a lot of energy to go deeper. Perhaps this turtle was fleeing from a shark.

References

'Rare marine reptile shatters dive record', *New Scientist*, 28 February 2004.

Guyton, Arthur C., 'Regulation of Respiration', *Textbook of Medical Physiology*, 9th edition, 1996, ISBN 0721667732, pp 525–535.

Iscoe, Steve and Fisher, Joseph A., 'Hyperoxia-Induced Hypocapnia: An Underappreciated Risk', *Chest*, July 2005, pp 430–433.

Phillips, Helen 'Into the Abyss', *New Scientist*, 31 March 2001.

Travis, John, 'Breathtaking Science', *Science News*, 4 January 2003, pp 8–10.

Dead Before You Hit the Ground

There's a fairly common belief that if you happen to fall from a great height, you'll be 'dead before you hit the ground'. The reasons given for this state include a heart attack, shock, fear of your imminent death, suffocation or generalised terror. On one hand, it is kind of reassuring to believe that you won't feel any pain if you fall from a great height. But the reality is that it's the huge deceleration as you suddenly stop that kills you.

Falling from a Really Great Height

It's really hard to die while you are in 'free fall', i.e. falling freely through the atmosphere. One scenario in which you can die in free fall is when you are so high up (say, above 100 000 feet or 30.5 km) that the intense cold and lack of oxygen will kill you. On 16 August 1961, US Air Force Captain Joe Kittinger rode a helium balloon to 102 800 ft (31 km), where the temperature was -79°C. It turned out that the air pressure was low enough at 62 000 ft (19 km) to boil the water in his blood (see 'Cold Tea and Boiling Blood' on page 81 for more information) — and at 102 800 ft (31 km), the

He hit the ground like a sack of ...

There is a common belief that if you happen to fall from a great height, you'll be 'dead before you hit the ground'. The reasons given include heart attack, shock, fear of imminent death, suffocation or generalised terror.

But it is the huge deceleration as you suddenly stop that kills you.

air pressure was a lot lower. He was kept alive by many layers of warm clothing; the thin, newly developed MC3 pressure suit covering his entire body; and a tank strapped to his body feeding him pure oxygen to breathe.

Captain Kittinger jumped out of his open gondola and began falling. By 90 000 ft (27.5 km) he had reached about 1149 kph — faster than the speed of sound. He was in free fall for about four and a half minutes. His speed gradually reduced to around 200 kph as he dropped though the increasingly thicker air, and his parachute opened at around 14 000 ft (4.3 km). There was a sudden jerk as his speed suddenly dropped to around 21 kph. He landed about 12 minutes later, with no permanent injuries, and he still holds two records: the only person to break the sound barrier without being in a craft, and the highest parachute jump. (For more information about Kittinger's feat, see 'Expanding Hand' on page 187.)

The important thing to realise from this account is that the act of falling freely does not kill you. You might become so terrified as you fall (which sounds very reasonable to me) that you psychologically 'freeze' and become unaware of your environment, but the act of falling does not kill you.

It's the G-forces

If you stop from 200 kph over a distance of a few centimetres, everything in your body suddenly weighs 7500 times more than normal. For example, your 1.5 kg brain briefly weighs 10 tonnes. In that brief instant, cells are burst open and blood vessels are torn asunder. The aorta (the huge main artery coming out of the heart) usually rips loose from the heart. For a few beats your heart continues to pump blood into the space around the heart and lungs, while no blood goes to your brain. But most of the blood vessels in your brain have also instantaneously torn loose. After that brief instant, your 'weight' returns to normal — but blood is now eating its way through your irreparably damaged brain.

There was a case of a parachutist who survived a fall (with broken limbs) when her parachute did not open. But the ground was very soft and, according to the fire officer, 'She left a good 12-inch hole in the ground.'

In general, most fit humans can just barely survive a sudden deceleration of about 40 Gs (where 1G is a force equal to gravity). In a 'typical' car crash at around 100 kph, the people inside would probably survive if they were wearing seat belts. The crush zone of the car is about 1.2 m — that's the amount by which the car shrinks. The airbag and/or seat belt slows you down over another 0.4 m. If the deceleration was even (which it is not), the victims would suffer decelerations of around 25–30 Gs. But the deceleration is usually not uniform, so there are very brief peak decelerations of 40–60 Gs.

So it's not the fall that kills you, but the Big Crunch at the end.

Cold Tea and Boiling Blood

A liquid such as water is full of tiny individual moving molecules. Some of these molecules are moving quickly, and some are moving slowly. These molecules have an average speed. There is a direct link between the average speed of the molecules and the temperature of the water. So, even on a cold day, a few of the molecules are moving rapidly. If they are near the surface, they can escape from the body of water into the atmosphere. This is how a puddle of water can evaporate, even in cold weather.

As you heat the water, the temperature rises, and so does the average speed of the individual molecules. The more heat you apply, the greater the number of molecules that are moving fast enough to escape from the liquid. These molecules will exert pressure on the surrounding atmosphere. Finally, you apply so much heat that the pressure exerted by these molecules is equal to the pressure of the atmosphere above the surface of the liquid. The water molecules now have enough energy to freely escape into the atmosphere. This is called 'boiling'.

At high altitude, the air pressure is low. A liquid will boil sooner at this lower pressure, which means that it has to be heated to a lower temperature than at ground level. At ground level, water boils at 100°C. At 4 km above the ground, the boiling point of water is only about 80°C. Blood is a liquid that is mostly water. In the body, its temperature is about 37°C. The boiling temperature of water is 37°C at 62 000 ft (19 km), so blood will literally boil at 62 000 ft — but only if you have it in an open container at 37°C. The elastic contractile pressure of the blood vessels adds enough pressure to the blood to stop bubbles from forming, so it won't boil in your body at 62 000 ft.

Reference
'Parachutist survives 2800 m fall', *Sydney Morning Herald*, 22 May 1991, p 11.

Kirlian Aura Photography

If you wander through a psychic fair, or a mind/body/soul festival, you will soon see somebody demonstrating Kirlian Photography. The photos show your 'aura', supposedly a representation of your spiritual and emotional self, or 'karmic field', or 'life force'. If you pay these people money, they'll sell you a colour photo of your aura. Unfortunately, the aura they sell is much more mundane than a real colour photo of your soul.

Human Aura

The human body definitely gives off heat, smells and sounds — but what is this 'aura'?

Well, in the Land of New Age, every living creature (whether it's a carrot, human or mosquito) supposedly has an aura that reflects the supernatural 'energy field' that, we are told, permeates all life. After all, don't depictions of Jesus Christ and various saints show a spiritual 'halo' surrounding the head?

In 1911, Dr Walter J. Kilner tried to bring together science, medicine and The Aura. He claimed in his book *The Human*

Atmosphere that he could see the human aura and use it to diagnose diseases. However, his diagnostic accuracy was pretty poor. The *British Medical Journal* was distinctly unimpressed.

The Kirlians Arrive

The next burst of aura interest occurred thanks to Semyon Davidovich Kirlian, an electrical technician in Krasnodar, Russia. In 1939 he noticed that if he touched high-voltage electronics in a darkened environment, a fuzzy, coloured glow extended around his body. He and his biologist wife, Valentina, spent years studying this strange phenomenon together. They passed high-voltage, high-frequency electricity through a test object while it sat on photographic paper. The current was kept very low, to keep the experiment relatively safe. When they processed the photographic paper, they could see where the test object had been — and a strange aura was left around the test object. This is what is known today as 'Kirlian photography'.

Kirlian believed that not only could he photograph auras but, more importantly, he could extract important diagnostic health information about plants, animals and humans from these photographs.

In 1961 the Kirlians published their results in the *Russian Journal of Scientific and Applied Photography*.

Kirlian Goes Mainstream

In the 1970s in the West, in the afterglow of Flower Power, some of the population were hanging out for cosmic stuff like auras. (I was one of those people — I was ready to believe.)

In 1971, Sheila Ostrander and Lynn Schroeder published their book, *Psychic Discoveries Behind the Iron Curtain*, and that's when Kirlian photography reached the West. The parapsychologist Thelma Moss also helped popularise Kirlian photography as a potential medical diagnostic tool with her books. She wrote *The*

Probability of the Impossible: Scientific Discoveries and Explorations of the Psychic World in 1974, and followed it up with *The Body Electric: A Personal Journal into the Mysteries of Parapsychological Research, Bioenergy and Kirlian Photography* in 1980.

You can still get Kirlian photography done with film and metal plates; the film is usually Polaroid film. But the more modern variation on Kirlian photography uses sophisticated scientific apparatus (okay, it's really just a video camera, computer and printer) to deliver the final image.

You don't need years, weeks or even days of training to be a Kirlian photographer, as it's incredibly easy to interpret one of these photographs. All you have to remember is that each colour stands for a mood: blue is peaceful and contemplative, green is healing or touching, red is vibrant or passionate, and so on.

It is also claimed that Kirlian images show nutritional deficiencies, psychiatric illnesses and even drug abuse. Kirlian photography also goes under the fancy name of 'bioelectrography' or 'bioresonance electrography'.

It's Just Sweat

So what does Kirlian photography really show? Is it the human 'bioplasma energy field' or our 'life force'?

Nope — a Kirlian photograph shows the presence of plain old water, or sweat, via a 'corona plasma discharge'.

The essential component to Kirlian photography is the high-voltage, high-frequency electricity. This travels really well through water vapour, and really poorly through dry air. Luckily for the photographers, people in an excited state sweat more. And the sweat turns into water vapour, which carries the high-voltage, high-frequency electricity more easily through the air.

This high-voltage, high-frequency electricity has enough energy to rip the electrons off atoms; atoms that have lost or gained an electron or two are called ions. So the air around the object becomes ionised, and if the air contains any water, you get this

You're lit up like a xmas tree

Photos of your aura supposedly show a representation of your spiritual and emotional self, or 'Karmic field', or 'life force'.

lovely glow. The scientists call this glow corona plasma discharge, but you can also think of it as 'slow lightning'. You see this effect in the sparks that come off your fingers in winter after you walk on a synthetic carpet, or in St Elmo's fire. (St Elmo's fire is a glowing cloud of hot, ionised gas that usually appears during thunderstorms, when there is a lot of electricity about. You can see it at a sharp point that is attached to the ground or water, such as a sailing ship mast, church steeple, or even the horns of cattle.)

And if you were in any doubt about the role of moisture in Kirlian photographs, the operator's manual for one commercial Kirlian photography device says:

'Photographing fingertips using the CV5500 Polaroid Conversion

1. The person being photographed first must MOISTEN THE FINGERTIPS of one hand with hand lotion or water, and place them FIRMLY on the metal touch plate that is on the top of the unit.'

The Actual Picture

Every Kirlian operator claims that the Kirlian image is unique to you. But if you go to five operators, and spend the money, you'll get five different images.

At least 25 factors influence the final image. These include the force and angle with which your hand is held against the metal touch plate, the atmospheric pressure, the voltage pulse rate, recent physical activity and, of course, humidity. Indeed, Kirlian photographs of the same person taken only minutes apart are so different (thanks to changes in water vapour levels) that they look like they came from different people. In fact, the Kirlian image disappears in a vacuum, where there's no water vapour. If it were due to a fundamental paranormal energy field, it should survive.

Some of the fairground Kirlian operators have taken the process one stage further. They take your photo with a digital camera, and then add the pretty colours around your body's outline from an image manipulation program on a computer — they just make it up on the spot and don't even bother with the high-voltage, high-frequency electricity thingie. It's basically the modern electronic version of sticking your face in a cardboard cutout of a famous person and being photographed as that person.

The Bottom Line

In other words, Kirlian photography is just a charlatan's magic trick which uses fancy pseudo-scientific apparatus to trick people into believing that the operator has some kind of scientific credibility. This belief helps transfer money out of your wallet into the operator's wallet.

So here's an easy experiment to do, if you have some money to burn. Get a Kirlian photograph made of your hand, and then wet the tips of your little finger, middle finger and thumb with a little saliva — and see how different the second Kirlian photograph is.

The Sceptics

In the 1980s, James Randi (The Amazing Randi) offered US$100 000 in a TV challenge to anybody who could prove they could see an aura.

One psychic claimed that not only could she could see the auras of 20 people on a stage, she could even see these auras extending beyond their bodies. This means that if the people were hiding behind a narrow screen, she would still be able to see their auras — even if she couldn't see the people.

This was the basis of the test. The psychic selected ten people whose auras she could see clearly (she claimed). They were then taken offstage, and 20 free-standing screens were put on the stage. Some of the screens had her 'aura generator people' behind them, and some did not. She then had to choose ten of the screens as having people hiding behind them. She gave only four correct answers out of ten. Even if she had been guessing, statistically speaking she would have got five out of ten.

Human Emissions

The human body gives off lots of stuff.

Your smell tells others about the state of your fertility, immune system and gender, amongst other things. You also give off sounds. And, yes, you do give off radiation — even electromagnetic radiation.

First, if light lands on you, some is absorbed but most of it is re-radiated. Light is electromagnetic radiation. Second, you are constantly emitting infrared electromagnetic radiation, which most of us call heat. Third, your brain gives off weak magnetic fields from all the electrical activity in the nerves.

Heat, light and magnetic fields are non-ionising forms of electromagnetic radiation. And we also give off ionising radiation (the stuff that can cause cancers). This comes from the radioactive potassium-40 naturally present in our body. If you sleep close to another human for eight hours each day, you will increase your dose of natural background ionising radiation by about 40%.

Do we give off an 'energy field'? Definitely not. I can say this with absolute certainty because physicists tell me that energy cannot exist as a field. You can have an electric field, or a magnetic field — but never an energy field.

Phantom Leaf

One argument that is often trotted out to 'prove' that something psychic is picked up by a Kirlian photograph is the 'Phantom Leaf' phenomenon.

First, lay down a leaf and take a Kirlian photograph of it and its 'aura'. Then lift up the leaf, cut off a part of it, and lay it down exactly where it was before. The Kirlian photograph now shows a ghost image of the missing segment!

The explanation is simple: a fresh leaf transpires (or breathes out) water vapour, which condenses on whatever surface it happens to be resting. Luckily, the emulsion of photographic paper absorbs water very well. When you replace the cut leaf, there is still some moisture left over on the paper from when the leaf was whole. So when you turn on the high-voltage, high-frequency electricity, the remaining water vapour on the paper sets off the corona plasma discharge, and you can see where the missing section of the leaf used to be.

So this 'missing halo' is not a mystical bioenergy currently unknown to science — it's just leaf sweat.

But here's a question: suppose that there is some strange 'missing halo' energy coming from the removed section of the leaf — why can't we ever see the 'missing halo' energy from the rest of the tree from which the leaf was picked?

'Real' Kirlian Photography

There is one situation in which you can see 'real' auras around another person: if you have a rare form of synaesthesia called 'emotion-colour synaesthesia'.

Synaesthesia is complicated, but you can think of it as a cross-wiring of the senses which means that a sound might trigger you to see a colour, and seeing colours may trigger tastes in your mouth. It occurs in about one in every 2000 people, and it tends to run in families.

Dr Jamie Ward from the University College, London, described a case of emotion-colour synaesthesia in a 19-year-old woman. She experienced coloured auras projecting away from her acquaintances and friends, and speaking emotive words to her such as 'hate' or 'fear' would fill her entire field of view with vivid colours.

References

Nickell, Joe, 'Aura Photography: A Candid Shot — Brief Article', *Skeptical Inquirer*, May 2000.

Ostrander, Sheila and Schoeder, Lynn, *Psychic Discoveries Behind the Iron Curtain*, Bantam, New York, 1971, pp 200–213.

Ward, Jamie, 'Emotionally mediated synaesthesia', *Cognitive Neuropsychology*, October 2004.

Lie Detectors and Lying Eyes

Pinocchio was a puppet with a nasty condition: he had his own built-in lie detector — his nose would get longer each time he told a lie. Unfortunately, we humans don't come equipped with such an obvious lie detector. However, most people reckon they know how to pick a liar. They believe that if somebody lies, that person will glance away, because the liar cannot look an honest person in the face. This is one myth about lying.

Another myth involves that famous piece of police investigative technology called the 'lie detector'. It has been used to try to extract the truth from criminal suspects and to screen job applicants. The lie detector doesn't actually detect lies — it measures stress, and it is easily fooled. Today's lie detector has a couple of very major problems: it doesn't catch all the bad guys, and it can wrongly incriminate the innocent.

The History of the Lie Detector

The various religions are fairly consistent in condemning liars. One of the Christian ten commandments is 'You shall not bear false witness against your neighbour'. In the Hebrew scriptures, Proverbs 12:22 reads: 'Lying lips are an abomination to the Lord'. In the Koran, Surah 3:61 carries a request to Allah to curse liars.

The lie detector has its roots in ancient Chinese interrogation technology. Suspects would have their mouths filled with dry rice; the guilty person would supposedly get a dry mouth because they were so nervous, and then find it really hard to spit out the rice. You can see one problem here: what if the guilty person was so used to telling lies that the interrogation didn't cause them any stress at all? In that case, they wouldn't have a dry mouth and would get away with their crime. And what about the innocent people who were nervous simply because they were on trial?

The modern lie detector is the result of several people's endeavours. They each worked on the belief that if somebody told a lie and felt guilty about it, their body would show a reaction somewhere — for example, the liar's face might not change, but their heart rate might speed up.

One such inventor was the psychologist Max Wertheimer. He started off as a musician, drifted into studies on the philosophy of law, and then studied the psychology of courtroom testimony for his PhD, which he received in 1904 in Germany. As part of his studies he developed a simple lie detector.

In 1921, medical student John A. Larson from the University of California worked with the local police department to develop his version of a lie detector. His lie detector measured changes in blood pressure, heart rate and breathing rate. Because it could measure several physiological reactions or responses at the same time, it was called a 'polygraph' — from 'poly' meaning 'many' and 'graph' meaning 'to write'.

Wonder Woman

The creator of Wonder Woman, the cartoon heroine, also invented a lie detector. Wonder Woman — who left her secret island of Amazons to fight the Axis forces during World War II — had a golden lasso which, among other properties, was a sort of lie detector. When it was looped around somebody, they simply had to tell the truth.

The creator behind Wonder Woman was Dr William Moulton Marston, who gained his law degree in 1918 and his PhD in psychology from Harvard in 1921. In 1915 he invented a machine to measure blood pressure, and later incorporated it into his version of a lie detector. As an expert witness in a murder trial in 1923, he argued that lie detectors should be admissible in court — but he lost.

In 1938 he appeared in an ad for Gillette razor blades; he used his lie detector to 'prove' that Gillette razor blades caused fewer 'emotional disturbances' than competitors' razor blades.

Marston was a very colourful character. He married Elizabeth Holloway and had two children with her. One of his students, Olive Byrne, later moved in with them, and also had two children with him. They all lived happily together, and he and his wife legally adopted Olive's children. After his death in 1947, Olive and Elizabeth lived together until they died.

Olive wrote articles about Marston for *Family Circle*, using the name 'Olive Richard'. One article, called 'Don't Laugh at the Comics', was about his belief that girls needed a role model of a strong woman, that men would find true happiness only when they could find a strong woman to submit to, and that this would lead to a peaceful world run by women. He also said that comics, with their massive readership, had a responsibility to put out this message. The editor of DC Comics (then called Detective Comics), Max Charles Gaines, hired Marston to create such a comic book heroine.

William Moulton Marston combined his and Gaines's middle names to create the nom de plume of Charles Moulton, under which he wrote 'Suprema, the Wonder Woman'. The character's name was then shortened to 'Wonder Woman', who made her debut in December 1941 in *All Star Comics #8*. Within a year she had her own comic and was the most popular character in the DC stable.

Marston's legal training paid off when he wrote a reversion of rights clause into his contract with DC which stated that if DC failed to publish a Wonder Woman comic in any given month, all rights to Wonder Woman (including all monies and merchandising of the past, present and future) would go back to Marston or his heirs.

'Going on the Box'

The premise of the early lie detector was the belief that the rate of breathing, the heart rate and the blood pressure would suddenly increase when the guilty party was asked a probing question.

A fourth measurement, the galvanic skin response (GSR), was later added — it measured how well skin (which is usually dry) carried electricity. Dry skin is about 40 mV (thousandths of a volt) negative (as compared to inside the body, which has a positive charge) and doesn't carry electricity well (in other words, its electrical conductivity is low). But skin has some two million sweat glands, so when the suspect is under pressure, the sweat glands squirt out water loaded with conductive salts. Simple electrodes can easily measure this increase in skin conductivity. In plain English, the GSR measures fingertip sweat.

However, the lie detector test involves much more than just getting physically connected by various wires and tubes to the polygraph. In the trade, this part of the test is called 'going on the box', because of the metal box that the polygraph usually comes in.

Before suspects go on the box, they undergo the 'pre-test', which is an hour-long interview during which the suspect gives their side of the story, and the examiner sees how the suspect responds. Following this, the examiner draws up specific questions for the suspect's test.

During the 'in-test', ten or so questions are asked of the suspect — but only three or four of those questions are related to the current investigation. The other questions are 'control questions' such as 'Is your name ___?' or 'Did you ever steal

Whisper me some of them sweet little lies

The lie detector aims to measure: Heart Rate, Blood Pressure, Breathing Rate and your Galvanic Skin Response (a measurement of how sweaty you become during questioning).

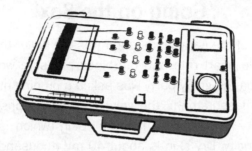

There is no such thing as a lie detector. The polygraph is designed to measure stress, which can be an indicator that someone is lying.

anything in your whole life?', both of which should get a positive response. Once the examiner knows how the suspect's body responds to fairly trivial questions, they can compare this to the suspect's response to stressful questions.

The final stage is the 'post-test', when the examiner analyses the readings to work out whether the suspect has lied.

Cheating the Test

One problem with the lie detector is that some people have no guilt in telling lies, so when they do tell a lie the needles of the polygraph don't swing wildly. Another problem is that any psychologist or decent 'spymaster' knows how to trick the polygraph. William Safire wrote in *The New York Times* about his 'old pal' Bill Casey, who was once chief of the CIA. Casey had challenged someone who accused him of skulduggery to take a lie detector test, and he would too. Safire 'wondered why he would take the gamble', and wrote that Casey 'reminded me that he was an old OSS (CIA precursor) spymaster, and that by using dodges like a sphincter-muscle trick and a Valium pill, he could defeat any polygraph operator'. Other tricks include biting on the tongue or counting down by seven from 100 whenever a control question was asked. This would throw off the baseline for control questions.

On the flip side, as many as half of the innocent suspects are so nervous that they register as guilty (that would be me).

If You Believe the Lie Detector ...

The lie detector is still used to screen potential employees for the CIA and other US federal organisations. The problem is that the CIA then believes that the employees who have passed need no further screening or follow-up tests — the lie detector gives a false sense of security.

One employee who definitely needed further screening was Aldrich Ames, who passed two separate lie detector tests for the CIA, even though he was actually a Soviet penetrator. Ames fed information to the Soviets, and his actions led to the executions of ten US agents in the then Soviet Union. This bears out William Safire's belief that the lie detector is a just a 'hit-and-miss machine measuring sweat, speedy heartbeat and other signs of nervousness [and] has been discredited as the judge of truth-telling'. Aldrich Ames did not receive a fabulous salary from the CIA, but he lived in a luxurious house and spent money like a drunken sailor. This anomaly would have been picked up by the most cursory security test. But the CIA believed that they did not need to check him out any further, because he had 'passed' the lie detector tests.

In 1987 a Cuban defector told the CIA that many double agents had passed lie detector tests to penetrate the organisation. In 1992 it was revealed that the Stasi, the East German secret police, had worked out how to trick the CIA lie detector and had also infiltrated the Agency.

There is another problem with the lie detector, too: it can ruin your life. Many Americans who have had to take a lie detector test to enter a branch of the US Federal Government (such as the Secret Service, or the FBI) have failed the test and yet claimed that the test was wrong. Once they have failed a single test, they cannot appeal against that test, they always have to reveal that they failed a lie detector test in any future job application, and they can never get employment in any federal agency that demands a lie detector test. Their entire future career is ruined — by somebody who has had ten weeks of training!

The Science

In October 2002, an expert panel set up by the National Research Council (part of the US National Academy of Sciences) delivered its report after 19 months of investigation. The report, called *The Polygraph and Lie Detection*, thoroughly rubbished the lie detector with the conclusion that 'national security is too important to be left to such a blunt instrument'. The report also said that 'no spy has ever been caught by using the polygraph'. The plain English version of their 398-page report is that the lie detector is rubbish.

On the other hand, the lie detector can sometimes trick a guilty and naive suspect into making a confession, simply because it is scary to see the needles of the lie detector jumping around.

According to the Pentagon, 90% of the information it gets comes via this untrustworthy pathway.

Lies and Eyes

Most people will tell a lie at least once per day, usually a little white lie along the lines of 'Your new hairstyle looks great'. And about two-thirds of us think that we can pick other people's lies by their shifty eyes, or when they refuse to look us in the eye — but we're wrong.

The Indian Vedas, ever practical, gave the advice that a liar will 'rub the big toe along the ground and shiver'. Unfortunately that advice is about as reliable as the warning about liars being unable to look us in the eye, according to Charles P. Bond, Professor of Psychology at Texas Christian University in Fort Worth.

Professor Bond carried out a study that involved over 2000 people in about 60 countries being asked the question: 'How can you tell when somebody is lying to you?' About 60%–70% of people (depending on the country) thought that the liar, overcome by the guilt of lying, would avert their gaze. But significant cultural differences affected the responses — for example, about 15% of people thought that liars would deliberately look you in the eye more often when lying, but this percentage varied between 30% for Muslims and 11% for Catholics.

Regrettably, there are no absolutely reliable ways to see if somebody is lying because there are huge differences in the behaviour of people who lie, and also in those who are lied to.

For example, most of us would feel some discomfort in telling a lie, and this discomfort would show itself in our facial expressions or body movements. But skilled long-term con artists got over this discomfort decades ago. And most of us are not skilled enough to pick up the micro-expressions that flicker across a liar's face as they change from one well-rehearsed mental mind-set to the next. These flickers of emotion last for less than one-fifth of a second — by comparison, your TV gives you 25 individual pictures each second. Just for fun, record some footage of your least-liked politician, watch their face frame by frame and feel free to interpret away!

Wizards Pick Liars

There are a few general behaviours displayed by liars: they tend to move their arms, hands and fingers less, and they tend to blink less. Because of the effort involved in keeping a story consistent, they tend to hold their lips tighter and have longer pauses in their speech, and sometimes their voices become higher pitched or more tense. They also have a habit of using fewer first-person pronouns and more negative 'emotion' words (such as 'hate' or 'bad'.) But these are all 'weak' signs.

Maureen O'Sullivan, Professor of Psychology at the University of San Francisco, surveyed over 13 000 people in an effort to find those individuals who can reliably identify liars. Because we are dealing with the incredibly complicated device called The Human Brain, and because there is deliberate deception involved, 'reliable' in this case means 'about 80% of the time'.

Most people were hopeless, but she had found some 31 people 'who we call wizards, who are usually able to tell whether the person is lying, whether the lie is about an opinion, how someone is feeling, or about a theft'. How? Well, it turns out that the 43 or so muscles in a face can make 3000 or so expressions that we might recognise. Professor O'Sullivan guessed that these wizards are very skilled at reading the momentary facial expressions called micro-expressions which are often too fleeting for non-wizards to pick up. These wizards included some lawyers (who are skilled with words), hunters (who have to look at their environment carefully), Secret Service agents (who guard politicians and have to scan crowds for non-verbal clues) and prison inmates (who are surrounded by liars).

But at this stage neither she nor the wizards really know how they do it. Nor do we understand how within Australian Customs it is the same group of officers who seem to be able to come at the top of the 'Contraband Recovered' list every month.

If only we were all like Pinocchio, and every time we told a lie, our nose got bigger — except, of course, when asked to judge somebody's hair, clothes, or driving skills ...

The USA Experience

Lie detectors are not widely used outside the United States of America.

The USA has about 3500 polygraph examiners, of whom 1500 do not belong to a professional organisation, such as the American Polygraph Association or the American Association of Police

Polygraphists. Each American state has its own laws regarding the licensing of polygraph examiners, and there is no central body in charge of this licensing. In general, the Associations require ten weeks of study and the passing of a written exam. (Ten weeks is not a very long time when you are dealing with something as sophisticated as the human brain.)

The US Federal Government uses polygraph tests more than any other organisation in that country. Use of the tests peaked in the 1980s, but began to drop after a review by the US Office of Technology Assessement claimed that it was impossible to truly measure how accurate the polygraph was because there was such a huge variation in the examinees, the examiners and the questions being asked.

How to Cheat

There are two main ways to cheat, i.e. to tell lies and not get caught.

One is to be so used to telling lies that telling more lies won't raise any stress in you. (Actually, this would not be a very pleasant way to live your life.)

The other way is to change your normal responses to telling lies. You could rub some antiperspirant on your fingertips so that you don't sweat, or you could try to generate the same extreme response to every question, so that your response to a lie is 'buried' in the 'noise'. So each time you are questioned, you could bite your tongue or press down on a thumb tack in your shoe. Alternatively you could dampen down all your responses by taking depressant drugs. Or you could try all those things at the same time …

Dr Charles Honts from the University of Utah found it was very easy to train people to cheat the lie detector using the method of creating stress whenever they were asked a control question. He asked some of his volunteers to 'steal' a valuable coin. Without training, one-fifth of his 'thieves' were found innocent by the lie detector; with training, this jumped to one-half.

A New Lie Detector?

Dr James Levine and his colleagues think that thermal imaging of the eyes might lead to a high-speed lie detector screening method. In their small study they asked eight volunteers to commit a mock crime, and then compared polygraph results with a measurement of the heat given from around the eyes when questioned. The accuracy of the polygraph was similar to the accuracy of thermal imaging. The sudden warming around the eyes is probably part of the flight/fright response, and probably as difficult to fake as normal breathing under stress, or a smile.

There is a difference between a real smile and a forced smile. In both of them you contract the zygomatic muscle, which runs from the corner of the mouth to the cheekbone, but in a real, spontaneous smile, you also contract the small muscles around the eyes. It is nearly impossible to contract these muscles voluntarily. On the other hand, with a lot of hard work, some people can learn how to control the muscles that twitch their ears.

So it may be possible to learn how to increase the blood flow to the muscles around the eye.

References
'The Word Liar', *New Scientist*, 29 March 2003, p 51.

Bonsor, Kevin, 'How Lie Detectors Work', *How Stuff Works* (www.howstuffworks.com)

Ekman, Paul, 'Lying And Deception', in Stein, N.L, Ornstein P.A., Tversky, B., & Brainerd, C. *Memory for Everyday and Emotional Events*, Lawrence Erlbaum Associates, Publishers, 1997.

Gaidos, Susan, 'Written all over your face', *New Scientist*, 12 March 2005, pp 39–41.

Koerner, Brendan I., 'Lie Detector Roulette', *Mother Jones*, Nov/Dec 2002.

Lock, Carrie, 'Deception Detection', Carrie Lock, *Science News*, Vol 166, No 5, 31 July 2004, p 72.

Pavlidis, Ioannis et al, 'Seeing through the face of deception', *Nature*, Vol 415, 3 January 2002, p 35.

Hail Caesarean Section

The 'normal' way for a baby to be born is via the vagina; the 'other' way is for the baby to be surgically removed. This other way involves cutting through the mother's skin, abdominal muscles and uterus. In surgery, 'to section' is to cut, so Australians tend to describe this operation using the phrase 'Caesarean section', while Americans tend to call it a 'C-section'. And sometimes it's simply called a 'Caesarean'.

If you ask most people, they will say that this operation gets its name from Julius Caesar, who was supposedly delivered by this method. Most historians, however, disagree.

The History of the Caesarean

The percentage of births by Caesarean varies around the world. It's about 15% in Sweden, 20%–25% in the USA and Australia, and up to 60% in Chile for mothers with private insurance.

But back in the days before sterile surgical procedures, antibiotics and anaesthetics, Caesareans were very uncommon because they were so risky — the mother would usually die,

Julius Caesar isn't the reason

Most people believe that the Caesarean section procedure gets its name from Julius Caesar, who was supposedly delivered by this method.

The how and where

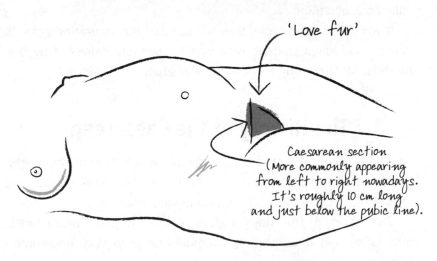

so the operation was only performed in the most extreme circumstances.

There are no mentions of successful Caesareans in the writings of Hippocrates, Aretxeus, Galen and Celsus. Even so, Greek mythology has it that Apollo delivered Asclepius (god of medicine) from his mother's abdomen. Caesarean-type procedures are also referred to in Hindu, Egyptian, Roman and other folklores. Pliny wrote that both Scipio Africanus and Manlius were delivered by Caesarean sections, and ancient Chinese etchings show Caesareans being carried out on women who apparently survived the procedure.

But the situation was very different in ancient Rome, the birthplace of Julius Caesar, with its poor medicine. Back then, Rome was expanding rapidly and desperately needed to increase its population, so they needed as many women as possible. Therefore, the abdominal delivery of a baby was performed only when the mother was dying or already dead. For in Rome it was very uncommon for a woman to survive the operation.

One of the first written records of a Caesarean in post-Roman times dates back to 1582. It claims that in 1500 a Swiss pig and cattle gelder, Jacob Nufer, successfully performed a Caesarean on his wife. Their baby would not come out, despite several days of trying and the help of some 13 midwives. Nufer got permission and 'surgically' delivered the baby, who lived to be 77 years old. Perhaps Nufer's work with animals had given him some surgical skills. His wife also survived and later gave birth to five more babies. However, the time delay between when the event supposedly happened and when it was written up makes it hard to believe.

In 1598, Jacques Guillimeau wrote a book on midwifery in which he introduces the word 'section' to replace 'operation'. Another source refers to a Caesarean section on a woman being done in 1610. Unfortunately, she died 25 days later, but this outcome was typical in those days. Even in the early 1800s, the maternal death rate from a Caesarean section was around 75%. The National Library of Medicine estimates that in Paris between 1787 and 1876, the death rate for women undergoing a Caesarean was 100%.

Man Who is Woman ...

The first recorded successful Caesarean in the British Empire was carried out by an army doctor ... who was actually a woman masquerading as a man. At some time between 1815 and 1821, a British Army physician, 'James' Miranda Stuart Barry, performed a Caesarean in South Africa.

In 1879 a British traveller, R.W. Felkin, saw several successful Caesareans carried out in Uganda by the locals. Banana wine was used in the operation and had a dual purpose: to 'anaethetise' the mother and to 'sterilise' the healer's hands and the mother's abdomen. The healer made a vertical cut along the midline and used a hot instrument to close the leaking blood vessels. The healer would also massage the woman's belly. Even then, the Ugandans knew enough about physiology to know that massage of the mother's uterus would make it contract, so shrinking its blood vessels and making it less likely to bleed. The uterus was not sewn shut where it was cut; instead, the two sides of the abdominal cut (in the abdominal wall) were brought together and closed with iron needles, and then covered with a paste made from roots. Usually, both mother and child would survive. Mr Felkin thought, from what he saw, that Ugandan medicine had been performing successful Caesareans for a long time.

Caesarean sections became safer for all with the introduction of anaesthesia and asepsis (working in a germ-free environment).

Theories

There are three popular ideas for how the term 'Caesarean' came to be linked with 'abdominal delivery of baby'.

First, when Julius Caesar ruled Rome, the law said that women in late pregnancy who were dead or dying should be cut open in order to deliver the baby. So this is one way that 'Caesar' might have formed the root of the word 'Caesarean'. Even if the state lost a fertile mother, it would at least get a living child. (In 1608, the Senate of Venice passed a similar law.)

The second theory is that 'Caesarean' comes from the Latin verb *caedare*, meaning 'to cut'. So perhaps this Latin verb led to the theory that Caesar was born by abdominal surgery, and then the procedure was given this name. However, Pliny says that Julius Caesar got his name from the Latin word for hair, *caesaries*, because he was born with a full head of hair.

The third theory is that 'Caesarean' comes from the Latin noun *caesones*, which refers to babies born alive after being cut from their dead mothers.

All this means that we can pretty definitely say that we're not sure how the name Caesarean came about.

And Caesar?

We do know that Julius Caesar's mother, Aurelia, was alive to hear of his invasion of Britain, so she must have survived his birth. Because the survival rates for mothers in ancient Rome were so poor, a Caesarean section was used only if the mother was dead, or about to die — and that makes it very unlikely that Caesar was born by a Caesarean delivery.

However, Julius Caesar did give us one special word: he was born in the month of *quintilis* — which was renamed July in his honour while he was still alive.

Shakespeare and Caesarean

Shakespeare, writing plays in the late 1500s and early 1600s, knew of Caesareans. His Scottish play, *Macbeth*, is a tragedy about Macbeth's driving ambition, and his rise and fall. It focuses on the worst crime of the day, regicide (the killing of a king), probably because of the interest raised in this crime by the Gunpowder Plot of 5 November

1605. The conspirators (English Roman Catholics) plotted to blow up Parliament and King James I, Queen Anne, and their oldest son.

In the play Macbeth, a general of King Duncan of Scotland, meets three witches. Their prophecies include that he will be king and that 'none of woman born' can harm him. There is a final battle between Macbeth and Macduff, during which Macbeth is killed by Macduff — who was not delivered vaginally, but was 'from his mother's womb untimely ripp'd'.

References

Gould, George M. and Pyle, Walter L., *Anomalies and Curiosities* of Medicine, W.B. Saunders & Company, Philadelphia, 1901, pp 128–132.

Sewell, Jane Eliot, *Caesarean Section: A Brief History* (A brochure to Accompany an Exhibition on the History of Caesarean Section at the National Library of Medicine, in Bethesda, MD, USA, 1993).

Encyclopaedia Britannica, Ultimate Reference Suite DVD, 2005.

Abreast of
the Times

Every now and then, people buying chicken breasts are amazed at how big they are compared with their size a few decades ago. And if somebody says, 'It's because of the hormones they put into the chickens', 80% of Australians will agree, because they believe that hormones are added to Australian chickens. But they're not.

Hormones

The word 'hormone' comes from the Greek *hormon*, which means to stir, rouse or excite. A hormone starts off as a chemical made by an organ; that chemical then circulates inside the body and affects other organs.

For example, the pituitary gland in the brain makes growth hormone that affects muscles and bones by breaking down fats to give energy, and uses that energy to make proteins. The thyroid gland in the neck makes thyroid hormone, which controls the oxygen consumption of cells around the body. The pancreas makes insulin, which has its effects in the fat, muscle and liver.

Back in the 1930s, it was found that if the mashed-up pituitary glands from cows were injected into other cows, they would produce more milk. It turned out that one of the many hormones in the pituitary glands was bovine growth hormone, which had the specific effect of increasing milk production. This hormone was later produced artificially, and in 1993 the American Food and Drug Administration (FDA) approved it for use in dairy cows. Today, about one-third of American cows are given this hormone, but no dairy cattle receive growth hormone in Australia. It is outlawed (only for marketing reasons, not health ones) and has never been registered for use in Australia.

Chickens and Hormones

Also back in the 1930s, the female sex hormone, oestrogen, was used on male chickens to 'chemically castrate' them, so making their flesh more tender. Under normal circumstances, if a rooster were to grow to full weight and maturity, its meat would be too tough. These chemically castrated roosters were called capons, and the technique was practised both in Australia and in the USA. However, it was discontinued in Australia in the early 1960s.

One hormone that was given to chickens and cattle — and to humans, with disastrous results — was DES (diethylstilbestrol), which acts as a synthetic oestrogen. It was first made in 1938, and American doctors prescribed DES to an estimated 5–10 million pregnant women to treat some problems of pregnancy, such as miscarriage. Beginning in the 1950s, DES was also used to make chickens and cattle grow faster. DES has been outlawed in the USA in chickens since 1959, and cattle since 1979, and is not used in any food-producing animals in Australia.

As it later turned out, DES did not help prevent miscarriages or early births. And in 1971 the FDA published a report that revealed the horrible truth: if a woman had taken DES directly (not indirectly from food she ate) while she was pregnant with a girl, that girl had an increased risk of vaginal cancer when she grew up. DES was immediately banned from use in humans.

Playing chicken with god!

Folk nowadays are constantly amazed at the size of the humble chicken breast compared to those of a few decades ago.

The simple fact is that NO hormones have been given to chooks for over 40 years.

Chicken breasts are bigger today due to the fact that chickens are bigger ...
Large chickens = large chicken breasts

Puerto Rico and Chickens

In July 1985 a TV documentary discussed a strange epidemic in Puerto Rico: young girls were advancing into puberty at a very early age. It was implied that this early puberty was being caused by hormones added to food; it was also implied that hormones were added to chickens around the world.

In follow-up scientific studies, the cause of the advanced puberty was never found. It was alleged that some Puerto Rican chickens had been given too many hormones, or hormones that were too strong, and that these chickens were accidentally fed to humans — this was never proven, though. Perhaps hormones were the cause, and perhaps they had entered the food chain via chickens — but perhaps not. Hormones were an easy target. There's no definite data in this particular case — the jury's still out.

Some data published in 2004 by the American Chemistry Council suggests that the culprit might have been phthalate esters, an industrial pollutant — and nothing to do with chickens. But, even today, we don't really know why this early puberty happened.

No Hormones in Australia

It's definitely true that hormones were fed to Australian chickens … in the past. Hormones have not been fed to Australian chickens for over four decades. Even so, 80% of Australians believe the myth that chickens are fed hormones, according to a survey released by the Australian Food and Grocery Council in September 2001.

Big Chicken Breasts

So why are chicken breasts so big today? Because the chickens are bigger.

These days chickens increase their weight by 5000% in five weeks, so that six-week-old chickens are four times bigger than they were 40 years ago. Dr Ian Godwin, a scientist at the Australian Poultry Research Centre at the University of New England in NSW, says that the rapid growth of today's chickens has nothing to do with hormones. It's 'entirely due to improved breeding programs, feeding regimes, and farm management'. And the Australian Poultry Research Centre is working towards reducing antibiotic use in chickens, improving their health, and reducing the environmental impact of chicken farms.

Life of Chickens

Just to kill another myth, chickens that are bred for eating do not live their lives only in wire cages but, rather, on the floors of large barns. These floors are covered with an absorbent material such as sawdust or rice hulls. The barns may not be the most pleasant places in the world, but they are not wire cages. There simply would not be enough cages to hold all chickens bred for eating in Australia: the birds have a 5–6 week growth cycle, and at any one time there are about 40–45 million broiler chickens in production in Australia, with eight million slaughtered each week.

Chickens that are bred to lay eggs, on the other hand, have a very restricted life. In most cases, they live in cages. They also usually live in unpleasant surroundings. Their lives are controlled rigidly to get them to lay one egg per day. They are brought up in an environment with about 17 hours of light per day, and are fed a diet rich in protein and fat. Laying chickens that deliver one egg each day are almost on the point of liver failure.

And yes, these chickens are occasionally fed antibiotics, mainly to control outbreaks of disease (such as necrotic enteritis). Only antibiotics that do not lead to resistance in bacteria that are likely to affect humans have been approved for use in the poultry industry. No antibiotic residues have ever been found in chicken meat in Australia, and the intention is to phase out the use of antibiotics over the coming years.

But chickens are not fed hormones.

References

Herbst, et. al., 'Clear cell adenocarcinoma of the vagina and cervix in girls: analysis of 170 registry cases', *American Journal of Obstetrics & Gynecology*, 1974, 119(5), pp 713–724.

Larriuz-Serrano M.C., Perez-Cardona C.M., Ramos-Valencia G., Bourdony C.J. 'Natural history and incidence of premature thelarche in Puerto Rican girls aged 6 months to 8 years diagnosed between 1990 and 1995', *Puerto Rico Health Sciences Journal*, March 2001, pp 8–13.

A Bird in the Hand

Most of us have a soft spot for cute little animals, especially the baby ones. But one thing that most of us hear or are told is that we should not touch a baby bird. The usual reason given is that the cute little birdie will be 'contaminated' by our human smell. The story goes that once the bird parents realise that their little baby bird (the nestling) is contaminated with the dreaded human smell, they cast it out to starve.

It turns out that there are sensible reasons not to touch or interfere with baby birds — but contaminating nestlings with our smell is not one of them.

Mess the Nest

One thing you should definitely not do is examine a bird's nest and then shift some of the eggs around, as some birds will think that a dangerous predator came to visit, and may still be lurking around. Upon making such a discovery, a few species of birds will abandon their nests, mostly permanently or, occasionally, temporarily. The eggs and resulting nestlings will be left helpless to fend for themselves. But note that, in cases like this, the parent birds are reacting to what they see — the shifted eggs. They are not reacting to the smell of the human who touched them.

In general, birds don't have a very highly developed sense of smell. Vision (to see food and predators) and hearing (to find subtleties of calls and songs) are usually the main senses of birds. There are exceptions. For example, the turkey vulture uses smell to find its dinner. But in general birds are not particularly offended by our human smell, and a human smell on nestlings does not condemn them to a short life span.

But let's look at the two types of cute little birdies: nestlings and fledglings.

Nestlings

The nestlings have nothing to do with a Certain Chocolate Company, but tend to be those birds who'd be in the neonatal unit. Nestlings have only a fuzzy down, not real feathers. They arrive in the world with a wide range of abilities.

At one end of the range, nestlings can be 'precocious'. These nestlings hatch fully feathered with their eyes open, and able to walk. Sometimes they even have some spare lunch in their bellies from the absorbed yolk in the egg. These nestlings (emus, brush turkeys, plovers, etc.) all move from the nest immediately and follow their parents. Some very precocious birds like brush turkeys never even see their parents, but dig their way out of their temperature-controlled mound and scurry off into the undergrowth. The north shore suburbs of Sydney are just getting used to this, as the brush turkeys invade their back yards.

At the other end of the spectrum are the 'altricial' birds, which are very similar to human babies — basically, helpless. These nestlings are born with their eyes closed. Because they don't have any feathers, they cannot regulate their own temperature when newly hatched. These are brooded in the nest by the parents until down and flight feathers appear, and they can perch. The parents then try to move them quickly onto branches to hop around so they can gain some strength.

The scent of a woman, or man ...

We are always told not to touch baby birds as they will become 'contaminated' with our human smell.

Birds usually don't have a strong sense of smell, so they won't reject a baby bird if it's touched by a human.

Most nestlings really belong in a nest because they can't take care of themselves.

There can be various reasons as to why they are out of the nest or abandoned in the nest: the parents may be inexperienced; the nestlings may have a very high load of parasites, or the nest may have been destroyed. Occasionally, a fellow nestling that is tougher than the others in the nest may have booted them out. This is relatively rare with herons, but more common with cuckoos such as the koel and channel-bills.

In most cases nestlings are found quite close to their nest (usually directly underneath), so the right thing to do is to put them back, and then their parents will start feeding them. If the nest has been destroyed, the best thing to do is contact a rescue and rehabilitation organisation such as WIRES (Wildlife Information and Rescue Service) in New South Wales, or Wildlife Victoria. If you have to take care of the nestling yourself, be ready for a busy day — some birds have to be fed every 10–15 minutes from sunrise to sunset. Swallows, for example, feed their chicks 500 times each day. And, of course, they need special food, not just a few leftovers from your fridge.

The food requirements of virtually all vertebrates (including birds) are almost the same: fat, protein, carbohydrates and calcium. However, the ratios vary from animal to animal, and at different times of their lives. This is regardless of what the animal becomes in later life — carnivore, vegetarian, or anything in between. If necessary, you can feed Farax, minced steak and Insectivore to nestlings. Insectivore is a product made specifically for the rehabilitation and aviary industry by two Australian companies, Wombaroo and Vetafarm. For the lay person, Insectivore is a strange product — it's basically the same as feeding the birds insects.

Fledglings

Fledglings have feathers and are slightly smaller versions of adult birds.

In general, fledglings don't need to be put back into a nest — in fact, being kicked out of the nest is part of their 'flight training'. They can't fly yet, so their parents drop in regularly to give them a feed until they mature and to teach them a few survival tricks. Juvenile cuckoos will beg food from their long-suffering parents for three months.

The best thing you can do for fledglings is to get them off the ground, leave them alone and keep your domestic animals away. For cats, bells worn on a collar don't really work — unless they weigh a few kilograms! But if the fledgling is injured (e.g. broken wing) contact your local rescue and rehabilitation organisation and follow their advice.

The THM (Take Home Message)

Birds usually don't have a strong sense of smell, so they won't hate a baby bird with human smells on it.

The first thing to do with a baby bird if you find one out of its nest is to work out if it's a nestling or a fledgling. In general, if it's a nestling, put it back in the nest. If it's a fledgling, it's just doing some flight training.

And if you don't know what to do, ask the professionals for some help.

Reference

Rescue, Rehabilitation and Release of Australian Birds, WIRES Training Manual.

Sleepwalking Awakening

When I was a kid, sleepwalking seemed terribly exotic, bizarre and strange. We (the kids in the neighbourhood and I) all thought that sleepwalkers stumbled around with their arms up in front of them, like Zombies. We were also all absolutely convinced that if you woke a sleepwalker you would kill them from the shock of the sudden awakening. But the opposite is true: if you wake a sleepwalker, you are more likely to protect them from death and injuries.

Sleepwalking Numbers

Sleepwalking covers a wide range of activities, from simple to elaborate. These range from merely sitting bolt upright in bed to the 'classic' sleepwalking, to boarding a train and travelling 160 km (as did 11-year-old Michael Dixon from Illinois in 1987), and even driving a car. Sleepwalkers can walk around, physically or verbally attack others, use the phone, cook and prepare food, wash their clothes or perform any other activities of daily living. Many sleepwalkers go back to bed to have a peaceful sleep for the rest of the night. Indeed, it's often the other family members who don't sleepwalk who have the most disturbed sleep, because they are woken by the sleepwalker. And sometimes, the sleepwalker may feel tired the next day.

Some people sleepwalk for up to an hour, but 5–15 minutes is far more common. Typically, the sleepwalking will start about an hour after going to sleep. Depending on your definition of sleepwalking, some 15% of all children sleepwalk. About 25%–30% of kids aged between five and twelve will do it at least once, and about 1%–6% will sleepwalk a lot. But, as the nervous system matures, the incidence of sleepwalking drops to about 2% in adults.

Definition of Sleepwalking

The medical name for sleepwalking is somnambulism, from the Latin *somnus*, meaning 'sleep', and *ambulare*, meaning 'walk'. Shakespeare describes Lady Macbeth as she sleepwalks: 'You see, her eyes are open, but their sense is shut!' Once again, Shakespeare was very perceptive.

The current psychiatric definition or diagnosis of sleepwalking requires six characteristics. These include: that the person (1) has repeated episodes of arising and walking about (usually in the first third of the sleep period); (2) has a blank, unresponsive face and can be woken only with great difficulty; (3) forgets all that happens during the sleepwalking; (4) is somewhat confused when they wake; (5) experienced some loss of functioning related to the sleepwalking; and (6) is not taking any medications that could cause sleepwalking.

Causes of Sleepwalking

We're not absolutely sure what causes sleepwalking, but it seems that it *may* result from an abnormality in the transition from deep sleep to the type of sleep you have when you dream — from non-REM sleep to REM sleep (REM is short for 'rapid eye movement'). If you look at the shut eyes of somebody who is dreaming, you can

Somnambulism

(From the Latin 'somnus' meaning 'sleep' and 'ambulare' meaning 'walk')

If you consider the damage that sleepwalkers can do to themselves, or to others, it's much better to carefully wake them and get them back to bed.

Sleepwalking can be triggered by many external factors including sleep deprivation, chaotic sleep schedules, general illness, stress and fever. Hormonal changes such as those that occur during puberty, pregnancy and menstruation can also set it off.

see behind their closed eyelids that their eyes are flicking to the right and left and back again, as though they are watching an invisible tennis match. This is REM sleep. When you dream, you are in REM sleep and you are usually 'paralysed' so that you can't act out your dreams (after all, if you did try to act out your dreams, you would almost certainly hurt yourself by running into walls). The abnormality present in sleepwalkers seems to be that when they drift into the dreaming state, they keep their muscle control and tone, and don't get paralysed.

Sleepwalking must have a genetic cause, as it can run in families. But it can also be triggered by many external factors, including sleep deprivation, chaotic sleep schedules, general illness, stress and fever. Hormonal changes such as those that occur during puberty, pregnancy and menstruation can also set it off.

Wild Walkers

There are several cases on record in which sleepwalkers have committed violent acts, and even killed a spouse or friend.

In 1878, Dr Yellowless described the case of a 28-year-old man with a long history of night terrors. He arose during the night and killed his 18-month-old son by smashing him against a wall. The man said that he thought he was fighting off a wild beast that was about to attack his family. He was found not guilty after the Lord Justice Clerk found that the man did not realise what he was doing because he was sleepwalking.

A similar case arose in 1933. Drs Howard and d'Orban describe a 31-year-old firefighter who 'woke to find himself battering his wife's head with a shovel. The shock was so great that he fainted and, when he realised his wife was dead, attempted suicide. He had no memory of getting out of bed, fetching the shovel, and there was an entire lack of motive, as he lived amicably with his wife.'

Back in 1985, Drs Oswald and Evan wrote about three cases in Scotland where sleepwalkers committed violent acts. In one case, a three-year-old boy was nearly stabbed to death by his 14-year-old cousin.

Dr E.B.L. Ovuga describes a sleepwalking murder in 1991 in Uganda, at a weekend residential health clinic. This was one of the few cases in which the incident was seen by witnesses. The 35-year-old man arose, walked into the next room and stabbed another man while loudly accusing that man of having raped him. He stabbed him to death, walked back to his bed and continued sleeping. The sleepwalker was kept indefinitely in a mental hospital 'for further observation'.

In general, violence in sleepwalking seems to involve males who are aged 27–48, and who have a strong family or childhood history of sleepwalking, bedwetting, nightmares and agitation on waking up. But while violence by sleepwalkers is bizarre and fascinating, don't forget that it is incredibly rare.

Wake a Walker …

If you wake up a sleepwalker, they will usually be confused and bewildered. Very occasionally, they will try to shake you off.

Sleepwalkers have performed very complex acts such as driving cars, sending text messages to friends, feeding pets, picking up random strangers and having sex with them, and so on. Certainly, you can hurt yourself when sleepwalking by walking off cliffs, onto railway tracks or just into household objects. If you have a sleepwalker in your family, sensible precautions include never letting them sleep in the top bunk, putting a bell on their door to warn you when they walk again, installing gates at the top of stairs and perhaps a lock on the fridge.

The Myth

So if you consider the damage that sleepwalkers can do to themselves or to others, you are definitely better off taking them gently by the elbow and leading them back to bed. And if they do wake up, you won't kill them by the sudden shock, or by a heart attack, as I and my young friends were once convinced.

Sleep and Eat

A truly bizarre method of self-harm while sleepwalking is SRED — Sleep-Related Eating Disorder. These people will arise while asleep, walk to their kitchen and pig out on food.

Their strange meals can include peanut butter wrapped in raw bacon, buttered cigarettes and cat-food sandwiches. Before being diagnosed, SRED sufferers could not account for their unexpected weight gain, nor for their cut fingers or scalded mouths (from drinking hot liquids). One sleepwalker luckily snapped into consciousness as she was about to drink corrosive ammonia cleaner.

You are doing these people a favour if you wake them.

Mowing the Lawn — Naked, at 2 a.m.

Rebekah Armstrong was woken at 2 am by a noise from her garden. She saw that her 34-year-old husband, Ian, was not in bed, and went downstairs. She describes what she saw: 'Ian was mowing the lawn completely starkers. I dread to think how long he'd been there but he'd nearly finished. I was going to wake him but I have always been told it can be dangerous to disturb someone who is sleepwalking. In the end I just unplugged the mower, went back to bed and let him get on with it.'

Ian later came to bed, but didn't believe Rebekah when she told him that he'd been mowing the lawn. Rebekah said, 'It wasn't until I told him to look at the soles of his feet that he finally believed me — they were filthy.'

References

Cartwright, Rosalind, 'Sleepwalking Violence: A Sleep Disorder, a Legal Dilemma, and a Psychological Challenge', *American Journal of Psychiatry*, July 2004, pp 1149–1158.

Howard, C, and d'Orban, P.T., 'Violence in sleep: medico-legal issues and two case reports', *Psychological Medicine*, 1987, Vol 17, pp 915–925.

Juan, Stephen, *The Odd Body 2: More Mysteries of our Weird and Wonderful Bodies Explained*, HarperCollins, Sydney, 2000, pp 239–244, 261.

Yellowless, D, 'Homicide by a somnambulist', *Journal of Mental Science*, 1878, Vol 24, pp 451–458.

Stomach Ulcers and Stress

At some stage of their lives, up to one in every ten Australians may suffer the pain of a stomach ulcer. Most of us 'know' that stomach ulcers are caused by stress or spicy foods, but two Australian medicos have overturned this commonly accepted myth. They've shown that an unusual bacteria causes practically all ulcers in the stomach and duodenum (the first bit of the small intestine).

Ulcer

An ulcer is simply a break in the lining of the local surface. So you can have ulcers on the skin, inside your mouth and, of course, inside your stomach.

The stomach carries acids that are strong enough to dissolve anything from meat to iron nails. The reason that your stomach doesn't dissolve itself is that a protective layer of mucus covers its inner surface and keeps the acid from touching the stomach lining. In addition, some of the lining cells produce bicarbonate, which neutralises any acid that comes near the lining. When something goes wrong and the acid does get access to the lining cells of the stomach, it may damage these cells and cause inflammation or ulceration. This sometimes leads to symptoms of a burning, or gnawing pain in the upper gut.

For a very long time, medical science didn't really understand stomach ulcers. Everyone knew that stomach acid was involved, so one popular treatment was drugs to reduce the amount of stomach acid. It worked while you took the medicine, but as soon as you stopped taking the drugs the ulcers came back.

Bacteria Surviving in Acid

It turned out that the vast majority of stomach ulcers are caused by a bacterium called *Helicobacter pylori*. But how can this bacterium survive in the acid of the stomach? It seems to be the only bacterium that can.

Its trick is to squirm through the layer of mucus on the inside of the stomach. The bacterium then sticks to the lining of the stomach, the epithelium, and manufactures an enzyme called urease, which has the special ability to create a tiny local alkaline environment around each individual bacterium. This alkali effectively neutralises the acid so that the *Helicobacter* can live in comfort.

This is such a clever trick that, even up to the 1970s, scientists couldn't believe that bacteria would survive in the incredibly acidic stomach.

Today, the diagnosis of *Helicobacter* can be made by looking for the presence of urease and there's a simple test of the breath to rapidly detect when *Helicobacter pylori* is present in the stomach.

History of Bacteria and Ulcers — 1

For the better part of a century, people were stumbling across the real cause of stomach ulcers ... but seeing is not believing.

In 1892, the Italian pathologist Giulio Bizzozero found some unusual spiral-shaped bacteria in the stomachs of dogs. He used a microscope with a magnification of 1000x and, to get clear images, used cedar oil to correct the light. His delicate and beautiful drawings got the shape and size of this strange bacteria exactly correct. But his findings were ignored.

The next close encounter with the truth was by the Irish professor Oliver Fitzgerald in the 1940s. He was a clinical pharmacologist with a special interest in duodenal ulcers. At that time the only treatments for stomach ulcers were a milk diet or surgery. The gut surgeons gave him pathology samples, and he noticed that ammonia and urea were usually present in the stomachs of patients with stomach ulcers. We now know that this ammonia and urea is part of the survival mechanism that *Helicobacter* uses to live in such an acidic environment, but Professor Fitzgerald thought that the ammonia and urea came from the stomach, not from any bacteria, because the belief of the day was that bacteria could not survive in the stomach. He believed this, and so he missed the chance to understand stomach ulcers.

My stomach is melting

One in 10 Australians will suffer pain from stomach ulcers at some stage in their lives.
It was generally thought that they were caused by stress or spicy foods.

Worry
(stress-related)

Hurry
(stress-related)

Curry
(hot and spicy foods)

BUT ... research has shown that unusual bacteria cause practically all stomach ulcers.

History of Bacteria and Ulcers — 2

In 1958, Dr John Likoudis, in the small Greek town of Missolonghi, about 100 km from Delphi, diagnosed himself as having a stomach ulcer. (This town is also famous for being the place where Lord Byron died.) Dr Likoudis treated himself with antibiotics, and got better. The antibiotics also seemed to work on his patients with stomach ulcers. He fine-tuned the mixture of antibiotics until he settled on four antibiotics, which he gave to his patients as a powder, wrapped in paper. His patients would take the powder for 7–10 days, apparently with good results.

Dr Likoudis patented his combination of antibiotics in 1961, calling it Elgaco; in 1962 he bought a pill-making machine to help him deal with the increasing number of patients. Over two decades, he treated some 30 000 patients with Elgaco.

Dr Likoudis was a lovely, kind man. For the eight years that he was mayor of Missolonghi, he gave his entire mayoral salary to one of the few pharmacists in town. Dr Likoudis would send his poor patients to that specific pharmacist, with instructions to say that the doctor said it was okay for the patient to get his medicines for free.

However, Dr Likoudis couldn't convince the medical establishment of the day that his treatment for ulcers was sound. In fact, he was fined for using an unconventional therapy. But he did publish a book in which he wrote, 'There is no doubt that gastritis and duodenitis which have gastric and duodenal ulcer as their complications, are inflammation due to an infectious agent'. He died, unrecognised by anyone except by his patients.

History of Bacteria and Ulcers — 3

In 1979, three doctors in Perth, conducting research on inflammation of the stomach, described bacteria in the duodena of some of their patients. They did not make a connection between this observation and ulcers, but they were working at the same hospital as two other doctors who finally brought the Bacteria-Cause-Stomach-Ulcers Theory into current medical thinking.

The two medicos working in the same hospital in Perth were Drs Warren and Marshall, and along the way they suffered rejection: their early papers were refused publication by both the American Medical Society and the Australian Society of Gastroenterology.

Robin Warren was a histopathologist with great powers of keen observation and a love of small cigars and strong coffee. He was fascinated by some strange bacteria he had found. Barry Marshall was a junior hospital gut doctor looking for a research project to boost his academic record. By 1982 they had realised that a new bacteria was somehow involved in causing stomach ulcers. They initially called this bacteri *Campylobacter pyloridis*, but when it was discovered that this bacterium did not belong to the Campylobacter family, it was renamed *Helicobacter pylori*.

It took many years before the medical profession became convinced that *Helicobacter* was the cause of stomach ulcers. Dr Warren had to hang onto his dogged belief that this new bacteria was important, and Dr Marshall needed enthusiasm and energy to keep their work moving forward. He also made the bold move (and sacrifice) of drinking a glass full of this bacteria. After a week he began suffering stomach pain, headache, nausea and vomiting, and hunger pangs even when full. Interestingly, though, he did not develop stomach ulcers, just gastritis. Different people have different responses to the same bacteria — obviously not everybody who has the bacteria develops ulcers, due to various factors. These include which version of the bacteria the person carries, how 'strong' their immune system is, and whether their genes make certain proteins.

By the 1990s, treatment of the bacteria as the cause of ulcers was becoming widespread. Even so, in 1995 in the USA, 90% of patients still blamed stress and spicy foods for causing stomach ulcers, and only 5% of medicos prescribed antibiotics to treat them.

Treatment

At some stage in their lives, about 40% of Westerners and 80% of people in poorer countries will become infected with this bacteria, but only about 20% of these people will get an ulcer during their lifetime. On the other hand, about 70% of people with stomach ulcers and 90% of people with duodenal ulcers are infected with *Helicobacter*.

Today, stomach ulcers are treated with combination therapy. Patients are given a drug to reduce stomach acids together with various mixtures of antibiotics. Combination therapy gives cure rates up to 85% — much better, and cheaper, than using acid-reducers alone. Not only do the ulcers heal, but getting rid of the bug usually stops the ulcer coming back.

So we now think that about 70%–80% of stomach ulcers are caused by *Helicobacter pylori*, about 20% are caused by non-steroidal anti-inflammatory medications, and about 10% are caused by a whole bunch of other factors (which may well include stress).

In short: hurry, curry and worry are not the sure-fire path to a stomach ulcer.

Drug Company Fortunes

In the 1980s, the then-current treatments for stomach ulcers mostly worked on reducing stomach acids, and were amongst the most profitable drugs on the planet. These drugs (such as Tagamet and Zantac) worked — after all, no acid means no ulcers. But as soon as the patient stopped taking the drugs, the ulcers came back. Of course, once the Bacteria-Cause-Stomach-Ulcers Theory came in, the pharmaceutical companies shifted over to treating stomach ulcers with antibiotics.

But things got distinctly rocky during the transition. For example, after Drs Marshall and Warren presented their work at a meeting in

Chicago, Glaxo shares dropped from $20 to $18 — the doctors' short presentation temporarily took about a billion dollars off the value of Glaxo.

References

Carlton, Sharon, 'The Helicobacter Story', *The Science Show*, ABC Radio National, 15 February 2003. http://www.abc.net.au/rn/science/ss/stories/s782213.htm

Fung, W.P., Papadimitriou, J.M., Matz, L.R. *Endoscopic, Histological and Ultrastructural Correlations in Chronic Gastritis*. *American Journal of Gastroenterology*, March 1979, pp 269–79.

Hamilton, Garry, 'Dead Man Walking', *New Scientist*, 11 August 2001, pp 31–33.

Lee, Professor Adrian, 'Helicobacter', *The Science Show*, ABC Radio National, 27 April 2002. http://www.abc.net.au/rn/science/ss/stories/s542387.htm

Are 'Natural' Remedies Always Safe?

Generally, people believe that 'natural' remedies — herbal medicine, vitamins and so on — are inherently safe and gentle, and have unique curative properties that conventional Western medications do not have. They are also Big Business. In the USA in 2001, consumers spent US$17.8 billion on 'dietary supplements', including US$4.2 billion on herbs and other botanical products.

According to US surveys, most Americans think that dietary supplements have three safeguards: they are approved and tested by a government agency; they have to carry warning labels about any possible side effects; and, finally, the manufacturers cannot make any claims about their products unless they are backed by scientific evidence.

Unfortunately for the consumer, none of these safeguards currently exist in the USA.

Drugs Have Side Effects

This situation is truly regrettable, because all drugs, including complementary and alternative medicines (CAMs), have side effects. Paracelsus (1493–1591), an early pharmacologist, wisely wrote, 'All drugs are poisons, what matters is the dose.' In other words, everything has side effects. (Yes, people have died from drinking too much water.)

For example, in the early 1990s a Belgian clinic treated some 10 000 patients with a weight-loss preparation that used a combination of Chinese herbs and Western medicines. In one of the batches, one herb (*Stephania tetrandra*) was replaced by a toxic herb (*Aristolochia fangchi*) — the Chinese characters for each herb appeared very similar to an untrained Western eye. By 1992, 70 dieters had complete kidney failure from this toxic herb while another 50 had severe rapidly progressive kidney failure.

By 1994, the herb had unleashed its second punch: any of the dieters who had taken more than a total of 200 g of the herb began getting cancers of the urinary tract. In 2001, the American Food and Drug Administration (FDA) issued a warning. Even so, in 2003 *Aristolochia fangchi* was still available on over 100 American websites, despite the fact that there were similar problems with this herb in France, Spain, Japan, the UK and Taiwan.

Other CAMs have their own special side effects. Comfrey can cause liver disease, yohimbe can give seizures and kidney failure, while ephedra (a cousin to the infamous 'speed' or amphetamine family) has killed hundreds of people from cardiovascular causes such as heart attack and stroke. Kava can cause liver failure, while St John's wort interferes with the metabolism of some other drugs and lowers the concentrations of indinavir (used in the treatment of AIDS) and cyclosporin (used in organ transplant patients). There have been cases of transplanted hearts being rejected by their new owners because those patients took St John's wort. Other CAMs can interfere with your blood's clotting reactions, and with anaesthetics.

The message is that CAMs can be very powerful, with pharmacologically active ingredients.

Of course, conventional Western medicines have their problems too. Thalidomide caused some 10 000 cases of pocomelia (seal-like limbs), and, more recently, Vioxx was linked with increased heart disease and stroke, and withdrawn from sale. Celebrex may be associated with an increased risk of heart problems, while Ibuprofen is linked to damage of the small intestine and heart problems. In fact, every prescribed medication, including aspirin and cough mixtures, will display a list of possible side effects.

Safeguards

One big difference between conventional medications and CAMs is the system that checks for side effects. Conventional medications are checked in two stages: they are tested before they are released, and then they are very closely monitored afterwards. This is how the warnings about Celebrex, Thalidomide, etc. came about. The drug company research may be biased, but there are other independent checks and balances.

In the USA, CAMs do not have to be checked at any stage. And the statistics are against you if you rely on the personal experience of just one single practitioner to identify potential side effects.

Suppose a dietary supplement causes a nasty side effect in one in every 1000 people. A practitioner would have to give this substance to 4800 patients, and then follow them up closely and often, to have more than a 95% chance of seeing this side effect at least twice. (If a practitioner sees a side effect just once, they may not link it to the medication, which is why this example is based on seeing the side effect at least twice.) That works out to giving the preparation to a different patient every working day for 18 years, or to two patients every working day for 9 years, and so on.

This is why, to accurately gauge the risks, we need a big database, which implies some central reporting body getting feedback on all medications given by all the practitioners.

So how did CAMs slip through the net in the USA? The answer seems to be lobbying by the manufacturers of CAMs.

Whoa ...it's just a herb, man!

There are those who believe that 'natural' remedies are all completely safe and gentle. Some folk also believe that natural 'herbs' have no adverse effects on their health or day-to-day functioning.

Do I need to ask ... whatcha lookin' at?

History of Drug Monitoring — 1906, USA

The first *Pure Food and Drug Act* was passed by the US Congress in 1906 and signed into law by Theodore Roosevelt. It was introduced because until then the totally unregulated market had allowed sawdust (and worse) to be added to food, and nasty drugs to be incorporated into pills.

Upton Sinclair's novel *The Jungle* described the very unhygienic conditions of the Chicago stockyards around that time. People were horrified about the extraordinary contaminants going into their food. Samuel Hopkins Adams wrote a series of articles about quack medicines in *Collier's* magazine in 1905; the public was outraged, and this led to the *Pure Food and Drug Act*.

This Act was only concerned with labelling, though. It required that all ingredients of foods and drugs, regardless of what they were, had to be accurately labelled on the produce. The consumer would then know if sawdust had been added to their sausages, or opium to their health tonic. If they then wanted to go ahead and consume that product, that was their business. But at that stage there were still no regulations about the products' effectiveness or safety.

History of Drug Monitoring — 1938, USA

The US *Federal Food, Drug and Cosmetics Act* of 1938 came into being because of a tragic poisoning caused by an elixir of sulphanilamide. It was marketed as a cherry-flavoured syrup, but it was dissolved in a solution of the toxic chemical ethylene glycol — and 107 children died.

In this case, the US government was able to stop the distribution of the elixir under the 1906 Act, because the product was sold as an elixir and the pharmacological definition of an elixir involves something dissolved in alcohol. Sulphanilamide does not dissolve easily in alcohol, so it was dissolved in ethylene glycol instead. Thus the manufacturers got busted for using the word 'elixir' but not using alcohol. This counted as inaccurate labelling, which was illegal under the 1906 Act.

By this time, though, the government realised that they had to go beyond merely accurate labelling. The 1938 Act looked at the safety aspect — that is, for the first time drugs had to be tested for safety. The aim of the Act was to make sure that drugs would not be allowed onto the market if they could kill you or made you sick. But, still, there was no requirement that they actually *worked*.

History of Drug Monitoring — 1962, USA

In 1962, the US Congress passed the *Kefauver-Harris Amendment* to the 1938 Act. This came about in response to the terrible side effects of Thalidomide. Finally, the drug companies had to prove that the drugs they sold would do what they were intended to do, and nothing else. Thalidomide is today being used to treat certain cancers, and it is now never given to pregnant women.

These three Acts (1906, 1938 and 1962) set the gold standard for drug control. Other countries all around the world looked to the USA for its excellence in the safety and effectiveness of drugs.

History of Drug Monitoring — 1994, USA

In regard to CAMs, however, these safeguards were lost when the US government passed the *Dietary Supplement Health and Education Act* (DSHEA) of 1994. This Act dealt with what was thought to be a new type of product — herbs and 'dietary supplements' that were 'neither drugs nor foods', but were 'thought' to have health benefits. They were now defined to contain: 'a vitamin; a mineral; an herb or other botanical; an amino acid; or a concentrate, metabolite extract, or combination of any of these ingredients'. They could be taken as a pill, a capsule, a tablet or a liquid, and they had to be labelled as dietary supplements. It is claimed that this Act was the result of effective lobbying by the health food industry.

The Act limited the roles and powers of the FDA, and deregulated the market. Dietary supplements would not be tested for safety, effectiveness or side effects; they were not subject to animal tests or clinical trials. A dietary supplement could be banned, but only after it had been proved unsafe by other means. Not surprisingly, sales of dietary supplements went through the roof.

Dietary supplements in the USA are now subject to lower standards than regular food supplements. One report wrote that 'consumers are provided with more information about the composition and nutritional value of a loaf of bread, than about the ingredients and potential hazards of botanical medicines'.

The Results

The US federal government did not even carry out post-marketing surveillance for the side effects of dietary supplements. However, American poison-control centres did, because poisoned people turned up for treatment. In 1998 these centres reported 6914 cases of adverse reactions specifically due to dietary supplements. There were 19 468 cases in 2001 and 24 412 cases in 2003.

The US Inspector General's report 'estimates that less than 1% of adverse events caused by dietary supplements, including herbs, are reported to FDA. Only a fraction of these are adequately investigated.'

For example, consider the product Metabolife 356, which at one stage was taken by 12 million Americans daily as a dietary supplement for weight loss. It was claimed to have 'the finest herbs gathered from over 12 countries of the world'. It didn't, but it did have massive amounts of caffeine and ephedra. It appeared on the market less than a year after the DSHEA was passed in 1994.

Metabolife 356 was devised by an ex-police officer, Michael Ellis, who had absolutely no medical training. He did, however, have lots of (illegal) pharmacological experience, as he had been busted in a raid on a methamphetamine laboratory in 1989. He plea-bargained his way out by agreeing to become an undercover FBI agent, supplying the Bureau with information on other drug dealers.

By 1999 Americans were swallowing 225 000 of Ellis's 'weight-loss' pills every hour. At one stage his company reportedly had an annual turnover greater than one billion dollars — but, over the years, it also received at least 13 000 complaints. Several hundred of the complainants ended up in hospital, many were seriously ill, and 100 died. These complaints were not reported to the FDA at the time because it was not compulsory to do so under the DSHEA. A class action suit is now being mounted against the company.

In Australia, when you take a prescribed medication you can be confident that the amounts of active drug are within a few per cent of the advertised levels. But this is not the case for CAMs.

Pan Pharmaceuticals was one of the largest manufacturers of CAMs in Australia, and it also made prescription medicines. It made a product called Travacalm, a regular registered medication, for motion sickness. But the active drug levels varied from 0% to 700%. The people who took the 0% pills had zero relief, while some of the 700-per-centers had wild hallucinations. If this was the level of care that Pan took with regular medications, what did it do with CAMs? In the same vein, the August 2005 edition of *Choice* reported that many of the over-the-counter preparations of glucosamine (for arthritis) had active levels down by as much as 28% from their advertised levels.

Where Now?

There are many similarities between CAMs and Western medicines: both are often based on natural products; both have good and bad effects; and both are marketed by Big Business. Maybe part of the reason why people believe that CAMs are totally safe is that the manufacturers do not have to advertise their side effects.

Many people use CAMs. To protect these people, I think that we need something like the German system, which has government regulation of the CAMs market.

Why Take CAMs?

There are many reasons why the market for CAMs is so huge.

CAMs are not usually taken when conventional medicine has an excellent cure, such as surgery for appendicitis. However, they are often taken when there is no real cure (e.g. for the common cold or an upper respiratory tract infection), or where the current treatments are not very effective (e.g. HIV, chronic intractable pain, some cancers, general unwellness).

Another reason is general dissatisfaction with mainstream medicine, or a desire to have more control over the treatment. This is helped by massive Internet marketing — after all, there is a huge amount of money to be made out of drugs, whether 'complementary' or 'Western'.

According to one survey, about 50% of people use CAMs. And about 50% of doctors don't ask their patients whether they use CAMs or not.

Problems with CAMs

There are a few potential problems with taking unpurified plant extracts such as herbs.

First, the plants can be wrongly identified, or they can be mispackaged.

Second, there can be different levels of active ingredients, whether because of a natural genetic variability in plants, variable growing conditions or different constituents (depending on what the herbalist mixes into that particular batch). Plants have thousands of different chemicals in them, and their proportions vary during the growing season, during the day, and so on.

Third, there can be many different contaminants, such as bacteria, viruses, fungi, parasites, microtoxins, pesticides, fumigating agents, toxic metals and even regular Western medications.

Consider the product PC-SPES, which was sold as a natural product to 'enhance prostate health'. It was found to be loaded with steroids, warfarin (used to kill rats), indomethacin (a non-steroidal anti-inflammatory agent), and so on. The company that made it, BotanicLab (of Brea, California), sold eight other 'herbal' products which were all laced with various prescription drugs.

In 1998 the California Department of Health found that 32% of 'Asian patent medicines' carried undeclared pharmaceutical drugs or dangerous levels of metals such as lead, mercury and arsenic.

CAMs don't even have to be supplied in child-resistant packaging.

'Natural' is Harmless

Quite often people have rung into my radio shows to ask about 'natural' products. Some of them have the belief that natural herbs are safe and gentle, and have no adverse effects on their health or daily functioning.

Marijuana is 100% natural. When I ask these callers if they would drive across Sydney Harbour Bridge in peak-hour traffic while under the influence of powerful marijuana, there is a long pause and then they reply along the lines of 'Oh, now I see what you mean'.

Competing Interests

I have worked as a hospital doctor at The Children's Hospital, Sydney, but no longer actively practise medicine. I also take CAMs occasionally, and would like them to be better monitored.

References

'Test: Glucosamine — Ease your Knees', *Choice*, August 2005, pp 24–26.

DeSmet, Peter, 'Herbal Remedies', *New England Journal of Medicine*, 19 December 2001, pp 2046–2056.Kessler, David A., 'Cancer and Herbs', *New England Journal of Medicine*, 8 June 2000, pp 1742–1743.

Marcus, Donald M., et al., 'Botanical Medicines — The Need for New Regulations', *New England Journal of Medicine*, 19 December 2001, pp 2073–2075.

Marcus, Donald M. et al., 'Do No Harm: Avoidance of Herbal Medicines During Pregnancy', *Obstetrics and Gynecology*, May 2005, pp 1119–1121.

Talalay, P., et al, 'The Importance of Using Scientific Principles in the Development of Medicinal Agents from Plants', *Academic Medicine*, March 2001, pp 238–247.

Not by a Long Sight (Galileo and the Telescope)

Astronomers have a special name for a telescope: they call it a 'light bucket'. You put out your light bucket at night and it catches a whole bunch of light from some distant object, and then you look at the light and see something wonderful. Indeed, the astronomer Johannes Kepler (1571–1630) said of the telescope, 'O telescope, instrument of much knowledge, more precious than any sceptre! Is not he who holds thee in his hand made King and Lord of the works of God?' Telescopes are as wonderful as they are cheap — you can buy a simple Galilean-style telescope for less than the cost of a bicycle. But most of us wrongly think that it was the famous Galileo who *invented* the telescope.

By the way, it is amazing to realise that over the four centuries we've had the telescope, we've been able to work out how the Universe began. That is a pretty remarkable accomplishment to spring from an instrument made from a few pieces of glass.

Ancient Lenses

One of the most famous ancient lenses is a piece of oval-shaped rock crystal, 42 mm by 34 mm across, and thicker in the middle than at the edge (6.2 mm vs 4.1 mm), which makes it a convex lens. It dates to the seventh century BC, and is called either the Nimrud Lens (from the ancient Assyrian capital in which it was found) or the Layard Lens (after its discoverer, the archaeologist Austen Henry Layard, who dug it up in 1849).

Many other convex lenses have been uncovered in the ancient world, in places such as Rome, Cairo, Carthage and Troy. Practically all of the lenses were steeply curved, and so had a short focal length. This means that they would bring their image to a focus only a few centimetres from the lens, so they were perfect for seeing fine detail close-up, especially if you were a bit older — but they were totally useless for a telescope. (It's actually much easier to accurately grind a steep curve than a shallow curve.)

Neither of the two lenses in a simple Galilean telescope is like an ancient convex lens, as the lens you need for close-up work is quite different from the lenses you need in a telescope. A basic, modern Galilean-style telescope has two quite different lenses: the light goes into the telescope through a shallowly curved, weak convex lens with a long focal length, and the light comes out into your eyeball through a steeply curved, strong concave lens (thicker at the edges than at the centre) with a short focal length.

'Inventors' of the Telescope

There are many contenders for the title of 'Inventor of the Telescope'.

One of the first is the brilliant Arabian astronomer and mathematician Ibn Al-Haytham (romanised to Alhazen or Alhacen). Around 1000 AD, he wrote about concave lenses in his ground-breaking mathematical book on vision (he also wrote on geometry, music, mathematics, politics and poetry). A few centuries later the Oxford scholar and Franciscan monk Roger Bacon (1214–1294) wrote about combining lenses to see at a distance.

Several others wrote about using lenses to see distant objects, including the Italian spectacle maker Giovanbaptista Della Porta (1538–1615).

But we have no evidence that any of them actually built such a device.

The 'Maker' of the Telescope

The first documented actual working telescope was made by Hans Lippershey, a spectacle-maker from Middelburg in Zeeland (south of Holland). He delivered this telescope to Prince Maurice of Nassau in The Hague; Prince Maurice was trying to negotiate peace with Ambrogio Spinola, the Spanish commander-in-chief in that part of the world, and the telescope had obvious military uses.

This event happened in the last week of September 1608, and the time was obviously ripe for the telescope to be invented because within three weeks, two other spectacle makers turned up in The Hague trying to sell what they thought was their unique new invention of the telescope. They were Zacharias Jansen, another spectacle-maker from Middelburg, and James Metius from Alkmaar (who was sometimes known as Jacob Adriaenszoon).

Galileo and the Telescope

The word about this new invention spread very rapidly across Europe. By May 1609, Galileo (then Professor of Mathematics at the University of Padua) had heard of this new device while in Venice. He was very smart and quickly worked out how to make his own telescope by thinking about refraction, the bending of light. He built his very first telescope the night after he returned to Padua, having taught himself all the necessary skills (including how to grind lenses). He soon made an 8-power telescope (which would make the image eight times larger), a 20-power telescope and a 38-power telescope.

Galileo quickly made enormous astronomical discoveries with these telescopes. For example, he proved that Ptolemy was wrong when he said that the Moon was perfectly smooth, as he could see mountains there — 'full of vast protuberances, deep chasms and sinuosities'.

Galileo also demonstrated that the Sun was not perfect, as he saw sunspots on its surface. This was another slap in the face for Ptolemy. And by tracking the sunspots, Galileo also roughly measured how quickly the Sun rotated.

Galileo showed that not everything in the Universe orbited around the Earth when he tracked the motions of Jupiter's four largest moons ... and showed that they orbited Jupiter, not the Earth. (Galileo cleverly called them the Medicean Stars, and was rewarded by the Medici family when they paid him to be their

Check out me 'light bucket'

It is a commonly held belief that Galileo invented the telescope. This is not the case. He refined the design of the telescope and, more importantly, used it to make some MAJOR discoveries.

Galileo: Overall smart dude. As an engineer he invented a thermometer, a compound microscope, a geometric military compass and even a machine for picking tomatoes.

personal philosopher and mathematician. Today's elite have their own personal trainers and financial advisers, but how many of them have their own personal philosopher and mathematician?)

He also observed the Sun's night-time shadow on the surface of Venus over time, otherwise known as the phases of Venus. You could see one series of night-time shadows on the surface of Venus if Venus went around the Sun, and a very different series of shadows if Venus orbited the Earth — so he proved that Venus orbited the Sun, not the Earth.

Galileo the Genius

Even today, simple telescopes are sold as Galilean-style telescopes. So how did Galileo get the reputation of inventing the telescope?

We're not really sure, but he was one of the first people to make great discoveries with the telescope. Perhaps the belief began when he presented his 8-power telescope to the Venetian Senate in June 1609.

But Galileo was also a towering intellect, so perhaps people automatically assumed that he had invented the telescope. After all, he had also performed basic scientific research into falling bodies and parabolic trajectories, and as an engineer he had invented a thermometer, a compound microscope, an escapement mechanism for pendulum clocks, a 'geometric and military compass' that helped gunners of cannons work out how much gunpowder to load and at what angle to elevate the cannon. He even designed a machine for picking tomatoes.

However, while Galileo made many great discoveries with the telescope, and had actually built his own from first principles, this all happened only after he had heard that other people had made the breakthrough. Galileo was, nevertheless, a man of far-reaching vision ...

Galileo and Light

Galileo was so much of a forward thinker that he actually tried to measure the speed of light. His experiment was doomed, though, because light travels very quickly — at 300 000 km/sec, or 300 m/microsecond.

Galileo arranged that his assistant and he would stand on nearby hilltops, each armed with a lantern equipped with a shutter that could open and close. Galileo would quickly open his shutter, and his assistant would open his own shutter as quickly as possible after seeing Galileo's lantern. Galileo would then measure the total time for the round trip — this was the time for the light to go from his hill to his assistant, the assistant's reaction time, the time for the light to return and, finally, his own reaction time.

The time for the round trip was the same whether Galileo and his assistant were a kilometre apart or a few metres apart. This was because the reaction time (say, one second) was much bigger than the one-way trip time (0.000003 seconds). Galileo realised that he would have to increase the distance between the hilltops enormously if he wanted to measure the speed of light.

References

Singh, Simon, *The Big Bang*, Harper Perennial, 2004, ISBN 0007152523, pp 3–37, 48–80.

Watson, Fred, *Stargazer: The Life and Times of the Telescope*, Griffin Press, South Australia, 2004, ISBN 1865086584, pp 37–83.

Encyclopaedia Britannica, Ultimate Reference Suite DVD, 2005.

Myths of Titanic Proportions

Even though the *Titanic* sank about a century ago, people still know of the story. There has been a resurgence of interest in the ship since the wreck of the *Titanic* was discovered sitting upright on the ocean floor nearly 4 km below the ocean's surface back in 1985, and a blockbuster movie was made about the disaster in 1997.

As a result, most of us 'know' that the *Titanic* was the largest and most technologically advanced ship of its day, and had indeed been advertised as 'unsinkable'. The captain was allegedly keen to claim the Blue Riband prize (for the fastest crossing of the Atlantic), and ordered full speed ahead for the ship's maiden crossing. And just before midnight on 14 April 1912, the *Titanic* smashed into an iceberg off Nova Scotia, a huge hole was gashed into her hull, and she sank. But that's practically all wrong.

Advanced Engineering?

In truth, the *Titanic* was quite conservative in its engineering, not advanced.

The *Titanic* (and her sister ships of the White Star Line, the *Olympic* and the *Britannic*) were more like lumbering 747s than speedy Concordes. They could carry huge numbers of paying passengers, get good revenue from each passenger, and were very large. Indeed, the *Titanic* was the largest ship of its day (fully laden, it weighed 66 000 tons). Showcases for advanced technology are, after all, expensive to build; instead, *Titanic* and her sister ships were deliberately designed with conservative technology, so they were cheaper to build.

However, *Titanic*'s competitors (the Cunard liners *Lusitania* and *Mauretania*) had more advanced engineering, even though they had entered service in 1907. Each of their four propellers or screws was powered by a high-pressure steam turbine, so they were faster. Each screw had a large balanced rudder, so they were more manoeuvrable. And each ship had won the Blue Riband.

By comparison, the *Titanic* and her sisters had conservative reciprocating expansion steam engines (running the two outside screws), with the excess steam feeding a low-pressure turbine which ran a third centre screw. Mr Bruce Ismay, chairman and managing director of the White Star Line, said, 'This method of

The Titanic: Some truths about the 'ride'

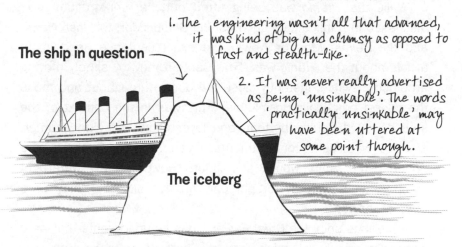

1. The engineering wasn't all that advanced, it was kind of big and clumsy as opposed to fast and stealth-like.

The ship in question

2. It was never really advertised as being 'unsinkable'. The words 'practically unsinkable' may have been uttered at some point though.

The iceberg

3. As far as an attempt at a speed record ... well, let's just say that Hollywood may have been in need of a little extra drama.

having two reciprocating engines with one turbine in the centre is more economical, as the turbine is fed by the exhaust steam from the reciprocating engines that would be otherwise wasted.' *Titanic* and her sisters had only one single tall slender rudder, making them less manoeuvrable than the Cunard liners (and thus less able to avoid running into large icebergs).

Titanic and her sisters, however, were a triumph for the accountants — three huge ships with cheap, conservative engineering. They were cheaper to build and run than smaller, more advanced ships.

Unsinkable?

The myth that the *Titanic* was advertised as 'unsinkable' is widely believed.

However, the phrase 'practically unsinkable' was used in the 1908 souvenir edition of *The Shipbuilder* in an advertisement for the Cunard ship *Mauretania*. The ad ran, 'Practically Unsinkable owing to the Watertight Bulkhead Doors being hydraulically controlled by the Stone-Lloyd System'.

While the *Titanic* was being fitted out, her construction was discussed in publications such as *The Shipbuilder*, the *Irish News*, and the *Belfast Morning News*. The exact sentence used in an article about the *Titanic* was, 'The Captain may, by simply moving an electric switch, instantly close the doors throughout and make the vessel practically unsinkable.' This was in the context of the ship having a double bottom (one layer of steel inside another) and 16 watertight compartments. Even if four of these were flooded, the ship would still float. However, the watertight compartments were watertight only on five sides — they were open at the top.

A previous ship, the *Great Eastern*, had compartments that were watertight on all six sides, but this sort of structure was very expensive to build. The *Great Eastern* also had a full double hull all the way to the waterline — again, very expensive to build. The

Titanic did have an extra hull layer, but only on the bottom, not on the sides as well. It was the full double hull that had enabled the *Great Eastern* to survive an 1862 encounter with a rock that opened up a hole 83 ft (25 m) long and 9 ft (2.75 m) wide in her side.

The White Star Line promotions never claimed that the *Titanic* was 'unsinkable' — the slogan they used was 'Largest and Finest Steamer in the World'.

Speed Record?

The myth also goes that, on her maiden voyage, the *Titanic* was speeding to New York in order to win the coveted Blue Riband for the fastest crossing of the Atlantic.

This is definitely wrong.

First, the *Titanic* was travelling on the longer southern route — to avoid the icebergs — rather than on the shorter northern route. If the captain had wanted to win the Blue Riband, he would have chosen the shorter northern route.

Second, not all of her boilers had been lit.

Third, even with all boilers lit, her maximum speed was 21 knots — much slower than the 26 knots of the Cunard ships.

Fourth, there had already been much correspondence within the White Star Line Company emphasising that the *Titanic* would berth in New York on a Wednesday (rather than on a Tuesday). This was to avoid the risk of damaging the engines, and the cost and inconvenience of taking care of passengers with fixed train or hotel bookings if they arrived early. It didn't seem as though they were in that much of a hurry.

Huge Hole?

Finally, the iceberg did not tear open a huge hole in the *Titanic* that reached a reported 100 m in length along her hull.

Robert Ballard led the team that found the *Titanic* sitting on the ocean floor. They explored and documented the wreck with several remote-operated vehicles as 'flying eyes'. They found little damage to the hull — six thin gashes in all, and the total area of these six holes was about 1.1 m^2.

The *Titanic* sank because of the location of these holes, not their size. The six holes were located over six of the sixteen watertight holds — and the ship could float with a maximum of only four of them flooded.

Icebergs

Each year about 10 000 to 15 000 icebergs break off from the Greenland glacier, but only about 500 make it to the North Atlantic shipping lanes. On this journey in 1912, there had been warnings about icebergs. However, the *Titanic*'s Captain Smith did three foolish things.

First, he ignored ice warnings.

Second, he refused to slow his ship after reports of icebergs in his direct path. The single slender rudder (part of the conservative technology) made the *Titanic* much slower to manoeuvre — and so it hit one of the icebergs in its direct path.

Third, he unnecessarily condemned an extra 457 people to drown in the icy waters because after the crash he allowed lifeboats to leave before they were fully laden. On that night, the conditions in the Atlantic were flat and calm. But the first lifeboat to leave carried only 12 people when it could have carried 40. The UK Board of Trade had required that the *Titanic* have lifeboat accommodation for 962 people, even though the ship could carry 3511 passengers. On its own initiative, White Line added four collapsible boats which could carry an additional 216 people between them, bringing the total lifeboat capacity to 1178. But only 721 people got on the boats. On that first and only voyage, *Titanic* carried 2224 passengers and crew, of whom 1503 died.

The One Truth

The only record that the *Titanic* did hold was that of being the largest ship in the world — and that soon sank without a trace after only five weeks on 23 May 1912, when the Hamburg-America line launched the *Imperator*.

How Did the Myths Start?

When the *Titanic* sank, the Hearst newspapers blamed Bruce Ismay personally. William Randolf Hearst, the newspaper owner, was at that time one of the most powerful people in America.

Hearst had met Ismay 20 years before the latter became chairman of White Star Lines. They disliked each other immediately and couldn't do business together, so intense animosity built up between them.

After the *Titanic* sank, the Hearst press ran a vicious campaign against the White Star Line and, in particular, J Bruce Ismay, calling him 'J Brute Ismay'. Ismay was accused of cowardice, and that 'he saved his own skin', and of leaving the sinking *Titanic* on the first available lifeboat. In fact, he helped others onto lifeboats and got into a lifeboat only while it was being lowered, and when there were no other passengers nearby. Later, at the British enquiry into the *Titanic* disaster, Lord Mersey wrote, 'Mr Ismay, after rendering assistance to many passengers, found "C" collapsible, the last boat on the starboard side, actually being lowered. No other people were there at the time. There was room for him and he jumped in. Had he not jumped in, he would simply have added one more life, namely his own, to the number of those lost.'

References
Titanic (the movie), 1997, Paramount Studios.
'Wreck Reckoning', *The Truth About History*, Readers Digest Association Limited, London, 2004, p 113.
Louden-Brown, Paul, 'Titanic: Sinking the Myths', BBC Online History, 1 April 2002.
http://www.bbc.co.uk/history/society_culture/society/titanic_01.shtml

Columbus and America

The schoolkid rhyme runs, 'In fourteen hundred and ninety two, Columbus sailed the ocean blue'. And that much is true — but he did not discover North America, he was not trying to prove that the Earth was round, and Queen Isabella of Spain did not pawn her jewels to finance his expedition. In fact, Columbus never set foot in North America.

North America, Ho!

If you want to get really picky, North America had been 'discovered' previously by people at least 20 000 years earlier. Even so, the Genoese sailor Christopher Columbus was not even the first European to 'encounter' the North American continent. That honour popularly belongs to Leif Ericsson and his Scandinavian colleagues in around 1000 AD (although there are claims that Bjarni Hergolfsson landed in Newfoundland in 986).

Christopher Columbus's voyages west across the Atlantic Ocean in the late 15th century were propelled by many forces. They included, on the part of Spain, his sponsor state, fear of both

Islam and Portugal; lust for gold and adventure; and, of course, the desire for spices.

At that time spices were used as medicines, and for flavouring and preserving food. They were a very lucrative trade because everybody wanted them (so the price was high), and they were easy to transport (being both small and light). But the Ottoman Empire and other Islamic states had made getting spices from the East very difficult, whether you went by land or by sea.

However, an alternative to dealing with the Middle Eastern states was the long-discussed 'long way around the world' to the East, accessed by first going West.

The World is Round

All the astronomers, astrologers and master mariners knew that the Earth was a sphere. After all, when a ship came over the horizon you saw the tip of the mast first and the hull last — exactly what you would expect on a spherical planet. Indeed, way back around 250 BC, the Greek poet, astronomer and scientist Eratosthenes had measured the circumference of the Earth within about 10% accuracy. In the early seventh century in Spain, the erudite St Isidore of Sevilla had published one of the first encyclopaediae, *Etymologiarum sive originum libri XX* or *Twenty Books on Origins, or Etymologies*. In it, St Isidore clearly wrote that the Earth was spherical. As a result, educated Europeans were quite aware that the Earth was a ball.

Columbus was also familiar with the calculations of the second century AD scientist, Ptolemy, who had discussed travelling to China overland. Columbus had also read of the travels of Marco Polo in the 13th and 14th centuries. Columbus was convinced that both Ptolemy's and Marco Polo's writings could be interpreted to mean that China stretched a really long way around the back of the Earth — so a determined navigator could travel to China in the East by first going West.

In an effort to suck up to his royal sponsors, Columbus wrote a preface to his account of his first two journeys. Part of it runs

Near enough, good enough

A few things about Columbus ... He did not discover North America, he was not trying to prove the Earth was round and Queen Isabella did not pawn her jewels to finance his expedition.

Columbus

thus: '... and I saw the Moorish king come out of the gates of the city and kiss the royal hands of Your Highnesses ... and I should not go by land to the eastward, by which way it was the custom to go, but by way of the west, by which down to this day we do not know certainly that anyone has passed ... Your Highnesses commended me that, with a sufficient fleet, I should go to the said parts of India ...'

Columbus mounted four journeys, none of which reached North America. His first journey was the most successful (especially because he came back with some gold), so he was given more funding for his second journey. But while his later journeys involved discoveries, they also didn't bring back huge quantities of gold — so the funding dropped off.

Columbus's Four Journeys

Columbus's first journey of 1492 was funded by a consortium of bankers, the King and Queen of Spain (King Ferdinand II and Queen Isabella I) and himself. So, yes, Queen Isabella was involved in funding his expedition. But, no, she did not have to pawn her jewels to do it — that particular myth was started by Bartolomé de Las Casas, a 16th-century Spanish historian. In his first journey, Columbus took three ships: the *Niña*, the *Pinta* and the *Santa Maria*. He visited San Salvador in the Bahamas (which he was convinced was Japan), and Cuba (which he thought was China). If he had continued east, he would have landed in Florida. But he turned southeast instead, and went to Haiti. He called it Hispaniola, and was convinced it was the Biblical kingdom of Sheba. He also found gold — which meant he could keep his backers happy.

His second journey (1493) was far better funded, possibly because his first expedition had brought back gold, spices, parrots and prisoners from the lands he visited. He set off with at least 17 ships, holding cavalry, some 200 private investors and 1300 salaried men. He again visited Haiti and explored more of Cuba.

Columbus's third expedition left with six ships in 1498. After visiting Haiti, he explored southwards and set foot in Venezuela; he was convinced that Venezuela was the temperate lowlands of an 'Earthly Paradise' and he interpreted the enormous floods of fresh water from the huge rivers as the Rivers of Paradise running into the sea. Further, he claimed that it was obvious that vast stores of gold must surely be nearby.

His fourth expedition of 1502 had four ships and visited Haiti, Jamaica, Cuba, Honduras, Nicaragua and Panama. He was forced to beach his remaining two ships in Jamaica, and was a castaway for a year. During that time, he correctly predicted an eclipse of the Moon in order to impress the locals, and thus scared them into giving him food.

Columbus Day

Today, Americans celebrate 'Columbus Day' on the second Monday in October to remember the landing of Columbus in the New World on 12 October 1492. And he *did* land in the New World — but the New World includes some land masses other than just North America. In reality, he never set foot in North America. The closest he got was the island of San Salvador.

So, first, Columbus did not ever land in North America. Second, Columbus was not sailing across the Atlantic to prove that the Earth is round — educated people already knew that it was round. No: Columbus was trying to find a better trade route to bring back lucrative spices from Asia. Third, Queen Isabella did not pawn her jewels.

Christopher Columbus died in relative poverty in Valladolid, Spain, in 1506. And to the end of his days, he was convinced that he had sailed to the Indies and China.

Who Discovered America?

First, all of us came from a mass migration out of Africa about 65 000 years ago. In only a few thousand years, humans had reached Australia by following the coasts. Our ancestors reached the continent of North America about 20 000 years ago (but some recent discoveries may have pushed that back to 30 000 years ago).

A Roman artifact — a 5 cm-tall terracotta head of a bearded man dated to around 200 AD — has been found in the Toluca Valley, about 65 km west of Mexico City. But there's no archaeological evidence that the Romans discovered North America.

In 458 AD a Chinese Buddhist priest sailed away on a 40-year journey to spread the faith. There are tantalising hints that he reached Mexico, as he describes a tree he found as having fruit that '… resembles the pear, but is red: the bark is spun into cloth for dresses, and woven into brocade.' This could be the maguey plant (*agave Americana*) from Mexico. He also describes the land as being poor in the metal iron — which is unusual (as most of the world has some iron), but is indeed the case for Mexico. This claim is made in volume 231 of the *Great Chinese Encyclopaedia*, written by the historian Ma Tuan-Lin in the 13th century.

There are also claims that the huge Chinese treasure fleet of 317 ships which sailed in 1421 may have reached North America, but the evidence for this is very thin.

Why 'America', Not 'Columbia'?

How did the continents of North and South America get their names?

They are named after Amerigo Vespucci (1454–1512), whose career is somewhat controversial. He travelled as a navigator on two (or perhaps four) naval expeditions — one (1499–1500) for Spain, and one (1501–1502) for Portugal.

In 1507, the humanist Martin Waldseemüller reprinted Amerigo's voyages as *Quattuor Americi Navigationes* or *Four Voyages of Amerigo*. He also published his own pamphlet, *Cosmographiae introductio*. In it he wrote that perhaps the massive, recently discovered southern continent should be called '*ab Americo Inventore … quasi Americi terram sive Americam*', which can be translated as 'from Amerigo the discoverer … as if it were the land of Americus or America'.

The idea caught on, and the name of Amerigo/America was later applied to both continents.

References

'Bones Settle Row of Whereabouts of Columbus's Body', *Sydney Morning Herald*, 12 August 2004, p 9.

'Who Discovered America', *The Truth About History*, Readers Digest, ISBN 0276427513, pp 250–255.

Macaulay, V., et al., 'Single, Rapid Coastal Settlement of Asia Revealed by Analysis of Complete Mitochondrial Genomes', *Science*, 13 May 2005, pp 1034–1036.

Thangarag, K., et al., 'Reconstructing the Origin of Andaman Islanders', *Science*, 13 May 2005, p 996.

Encyclopaedia Britannica, Ultimate Reference Suite DVD, 2005.

The Great Cash Crash of 1929

There are certain dates in history that have a special significance, depending on your age and the culture in which you live. In more recent times, these might include 7 July 2005 (London bombings), Boxing Day 2004 (Indian Ocean tsunami), and 11 September 2001 (World Trade Towers, New York). The twenty-eighth of October, 1929 is a long way back in time. But if you have ever followed the stock market, you might recognise it as the day of the Great Stock Market Crash of 1929.

We probably 'know' two 'facts' about this Crash: first, that this huge Crash happened on one single day, plunging the USA and most of the industrialised world into the Great Depression; second, that the Crash was accompanied by a huge jump in the suicide rate. Unfortunately, both 'facts' are wrong.

Real Estate to Stock Market

Professor John Kenneth Galbraith gives a very nice summary of the Crash in his book *The Great Crash: 1929*. He describes how, early in the 1920s, there was a crazy real estate boom in Florida.

Crash ... I heard no crash!

If a stock market crashes in a forest, will anybody hear it?

28 October 1929 ... the day of the Great Stock Market Crash

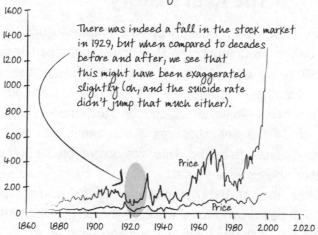

There was indeed a fall in the stock market in 1929, but when compared to decades before and after, we see that this might have been exaggerated slightly (oh, and the suicide rate didn't jump that much either).

Stock Prices and Earnings, 1871-2000

In a regular real estate boom, people buy and then sell the land, making a quick profit. But this crazy boom was so wild, and prices were moving so fast, that people sold not the actual real estate but just the right to buy and sell it. This right, called a 'binder', cost just 10% of the land's price.

This introduced into the minds of the American public the concept of 'buying' something without paying the full price. This mind-set was one of the reasons that interest in the stock market in the USA increased so hugely from the mid-1920s.

As a result, there was a 'shortage' of shares, so speculators set up investment companies that would sell shares in themselves, which increased the number of available shares listed on the stock market. However, these investment companies only bought shares in other companies, so in a sense they were built on sand.

Then there was the rise of 'boiler rooms', or 'bucket shops', that sold shares over the phone. (Yup, those cold calls on the phone from strangers trying to get you to invest in something have a long and dubious history of some 80 years.) As a result, by the late 1920s, 1.5 million of the 120 million people in the USA were actively involved in the stock market.

The Real History

The American stock market actually peaked in late August 1929, two months before 28 October. The Dow Jones Industrial Average closed at 380 at the end of August, and during September and October the prices began to drop — but still people were buying and selling in a frenzy. There was a sudden stock market drop on 18 October; on 24 October there was a real panic, with over 12 million shares traded — a record. This was beaten on 29 October, when 16 million shares were traded.

However, the market did not collapse to its lowest value overnight. It bumped around for a while — the Dow Jones closed at 274 at the end of October, but had jumped back up to 286 at the end of March 1930. It took three years of slow-motion tumbling for the market to bottom out, by which time the Great Depression was well under way. The market did not recover to the level of its August 1929 values until the 1950s.

So the stock market definitely did not fall all the way down in just one day.

Suicide Rate

So what about the massive numbers of suicides that the London penny press wrote about avidly?

In his book, Galbraith writes, 'Clerks in downtown hotels were said to be asking guests whether they wanted the room for sleeping or jumping' — but then he discounts this myth. He tabulates the suicide rates from 1925 to 1935, looking both at the entire USA and New York City. The suicide rates for both have a very similar trend, reaching their peaks in 1932 and then by 1934 dropping back to their 1929 levels.

How did this myth arise? Well, on Thursday, 24 October 1929, a workman appeared on a New York building top to make some repairs and a crowd quickly formed, expecting him to jump (which he didn't). This was written up, and contributed to the general hysteria. And any suicides that did happen were automatically blamed on the stock market crash.

Irrational Exuberance

William Feather (1889–1981), an American author and publisher, wrote very wisely about the irrational exuberance that created the events of 1929: 'One funny thing about the stock market is that every time one person buys, another sells, and they both think that they are astute.'

References

Galbraith, John Kenneth, *The Great Crash: 1929*, Houghton Mifflin Company, Boston, 1954 (reprinted 2004 as 50th Anniversary Edition), ISBN 039513935X.

Goodspeed, Bennet W, *The Tao Jones Averages: A Guide to Whole-Brained Investing*, E.P. Dutton, USA, 1983, ISBN 014117368X.

Schiller, Robert J., *Irrational Exuberance*, Princeton University Press, 2000, ISBN 090801158X.

Cleopatra: Beauty or Beast?

The world is fascinated by Cleopatra. Cleopatra (Cleopatra VII to be exact) was the last Pharaoh of Egypt, and has inspired books, plays, movies and 32 operas. Most of us are not experts in Egyptology, but we all think that we know a few things about Cleopatra — something along the lines that this Egyptian woman was stunningly beautiful, and committed suicide by getting a small snake, an asp, to bite her. The only correct belief in all of that was that she was a woman.

Cleopatra Egyptian?

Cleopatra was not Egyptian.

Alexander the Great (who was from Macedonia) conquered Egypt and after he died in 323 BC, control of that country passed to Ptolemy, the son of Lagus, a Macedonian nobleman. Ptolemy set up a three-century-long dynasty which was closed to outsiders, including Egyptians. Indeed, it was so exclusive that the parents of Cleopatra were brother and sister (a fairly common practice in Egyptian royal families since at least New Kingdom times). The *Encyclopaedia Britannica* writes that 'the Macedonian-Greek character of the monarchy was vigorously preserved' and that 'Cleopatra VII was of Macedonian descent and had no Egyptian blood'.

In fact, Cleopatra was the first of the Ptolemaic pharaohs to learn the Egyptian language. She was also the last pharaoh of Egypt — after her death, Egypt became a province of Rome.

Cleopatra Beautiful?

Cleopatra was almost certainly not beautiful in the physical sense (bearing in mind that the concept of beauty is different for each time period and each person on the planet).

The Greek essayist Plutarch wrote of Cleopatra about a century after her death in 30 BC, in his book *Life of Antony*. He describes her as 'by no means flawless or even remarkable' and writes that 'her beauty was not in and for itself incomparable'. Only ten coins from her reign displaying a representation of her have survived in good condition. They show her as having a fat neck (euphemistically called 'Rolls of Venus'), a hooked nose, long ears and a prominent chin. Cleopatra was, like all the other Ptolemaic women, around 1.5 m tall. So, in today's terms, she was short, dumpy and squat — but was also very appealing.

If you follow Frank Zappa's belief that 'your brain is your main sex organ', Cleopatra was beautiful. She could speak nine languages, and had a sharp intelligence and compelling charisma. She was highly educated, had a powerful regal presence from an early age, and projected a sense of vigorous charm and powerful leadership. Plutarch writes: '... her conversation had an irresistible charm ...' and 'the seduction of her speech ... her character, which pervaded her actions ... was utterly spellbinding. The sound of her voice was sweet.' She was so utterly charming, wrote Plutarch, that 'Plato admits four sorts of flattery, but she had a thousand'. Her wisdom and wit are praised in Arabic and Coptic literature.

Plain as a lemonade icy-pole!

It's becoming clear that Cleopatra may not have been the physical 'beauty' history has led us to believe.

The Uglee© test

☑ The 'Rolls of Venus" (Fat Neck)
☑ Hooked Nose
☑ Long Ears
☑ Prominent Chin
☑ Short (1.5 metres)

The Uglee© rating: 8.5 out of 10 ... short, dumpy and squat.

So while she was not physically beautiful, she had what it took to woo and win the hearts of two of the most powerful men of her day, Julius Caesar and Mark Antony. Julius Caesar said, 'No blood was in her veins, but the sun's blood. Sweet Hathor lived in her eye and her dimpled knees.'

Did an Asp Kill Cleopatra?

Shakespeare tells us in *Antony and Cleopatra* that Cleopatra died from the bite of an asp which had been smuggled into her bedroom in a basket of figs.

But that particular species of snake does not live in Egypt. A far more likely contender (if she was indeed killed by a snake) was the cobra, which does live in Egypt. Plutarch writes: '... but on opening the doors, they saw her stone dead, lying upon a bed of gold, set out in all her royal ornaments'. Even today, we still don't know how she really died.

An asp did appear as the 'uraeus' on her crown. The uraeus was a representation of the sacred serpent as an emblem of supreme power, and was normally worn on the headdress of Egyptian gods and pharaohs. Perhaps this asp on her crown prompted this myth?

Her Mummy Case

Unfortunately, the science of forensics will not be able to show us Cleopatra's face. Napoleon looted many treasures from France, including the mummy case of Cleopatra VII; most of these treasures were returned to Egypt, but her mummy case was accidentally left behind. In the 1940s, French workers found her mummy case, and emptied it into the sewers …

Cleopatra — Miracle Worker

One image that we don't have of Cleopatra is of her sitting at her desk, reading military intelligence and locust control reports.

And yet she must have, to have been able to bring back her kingdom from the edge of doom the way she did. She inherited a bankrupt kingdom that couldn't even afford to mint coins from gold, only from lesser metals (which is why we have so few coins with clear images of her).

She raised an army, built a fleet to rival Rome's, made Egypt strong, kept the peace, and successfully played off powerful opponents against each other. Her capital city, Alexandria, was probably the most sophisticated city in the world: it had a universal health service, autopsies, and a magnificent library and lighthouse, and was a centre that drew in artists, scientists, engineers and writers from across the world.

References

Did You Know, Readers Digest Association Limited, London, 1990, p. 249.

'Plain and Poker-Faced: Cleopatra's supposed beauty went unnoticed during her lifetime', *The Truth About History*, Readers Digest Association Limited, London, 2004, ISBN 0275427513, pp 288–293.

Walker, Susan, 'Cleopatra of Egypt: From History to Myth', *The New Republic*, 4 April 2002.

Encyclopaedia Britannica, Ultimate Reference Suite DVD, 2005.

Off Ya Banana Tree and Feeling Fruity

The banana is so popular in Uganda that the locals eat about 250 kg each per year. In fact, the banana is so essential to their diet that their word *matooke* means both 'food' and 'banana'. The banana is not quite as popular elsewhere, but kids around the world like it because it's tasty, and because it's so easy to peel and devour. And they believe two things about bananas that are wrong. First, that the yellow banana is not a fruit (because it has no seeds), and second, that it comes from a tree (the banana tree).

History of the Banana

There are about 500–1000 species of banana today. Most of them are inedible — they carry hard, pea-sized seeds and have only a small amount of bad-tasting flesh. The botanists think that about 10 000 years ago, probably in Southeast Asia, a random mutation produced a sterile banana with no seeds and lots of flesh that could be eaten uncooked. The internal dark lines and spots inside today's banana are the vestigial remnants of the original seeds.

Bananas were taken to India, where Alexander the Great saw them, and where they appear in 2500-year-old cave paintings. Traders took them from India to East Africa, then overland to West Africa. Portuguese sailors then transported them to the Canary Islands, from where they travelled to Haiti by the 15th century. They were imported into North America shortly after European settlement, and became freely available in American fruit markets by the 19th century.

Bananas and Australia

Bananas first came to Australia by two separate pathways, one for the west coast and one for the east coast. They arrived in Western Australia in the early 1800s when Chinese workers entered the country and carried some plants with them to Carnarvon.

Bananas travelled to Queensland, however, via a very convoluted pathway. In 1828, two banana plants were taken from Mauritius and hand delivered to Lord Cavendish in England. He grew some bananas from them in what became Kew Gardens. Missionaries took some of his banana plants to the South Pacific in 1840, where they flourished. Later, a missionary called Williams took some of these bananas to Fiji. In the 1870s, Queensland sugarcane plantation owners 'recruited' sugarcane cutters from Fiji and brought back some banana plants. In 1891, Herman Reich used these bananas to start the Coffs Harbour banana industry in New South Wales.

Life Cycle of the Banana

Even though the banana has a phallic shape, it is a sterile and mutant fruit that has not had sex for 10 000 years. So how do we get new bananas?

New banana 'trees' are 'born' in a new location when cuttings are planted in the ground and take root successfully. The cycle starts with an underground stem (called a rhizome), often measuring

The Banana: Is it a fruit or is it a vegetable?

Monkey-boy, a long-time fan of the banana, has never before questioned his most-loved food.

The ultimate (simple) fruit/veg test
If it goes with ice cream ... it's fruit
If it works with gravy ... it's a veggie

metres across, that can have several banana 'trees' growing from it. Each of these 'trees' started from an underground 'bud'.

A 'bud' will push up a shoot which breaks through the soil. The shoot is made of leaves wrapped tightly around each other, so that it looks like a green tree trunk, even though there is no wood present. The oldest leaves are on the outside, with the newest leaves pushing upwards through the middle.

When the time is right, the underground rhizome switches from making leaves to making an 'inflorescence', which is a flowering structure. The inflorescence has a broad leaf-like structure (called a bract) that wraps around a hand of flowers, which ultimately turns into a hand of bananas. The final 'tree' can be up to 6 m tall, with bunches of 50–150 individual fruits or 'fingers' of bananas, broken up into hands of 10–20 bananas each.

Once that particular underground bud has grown an inflorescence, it cannot reset itself to growing leaves and, ultimately, another 'trunk'. It has done its dash. So that bud and trunk will die and wither away. But the next summer, other buds will appear on the underground stem — and so the cycle continues.

The Truth

So bananas are definitely a fruit, even though the fruit is sterile and has no seeds.

And the banana hand does not grow on a tree: it grows on a plant. It's not a tree because it's made from leaves, not true woody tissue.

And why are cartoon characters in the movies and on TV always drawn with only four fingers? The standard answer that cartoon animators tell me is that if you give them five fingers, the hand looks like a bunch of bananas.

End of Bananas?

We have already 'lost' one species of banana. French botanists uncovered the Gros Michel in Asia in the 1820s; it was a rich, sweet banana which, if you chose to eat it while it was green, didn't have a bitter aftertaste. In the first half of the 20th century, the Gros Michel was virtually the only sweet banana sold in Europe and the USA. But it was very susceptible to a fungus in the soil which caused an incurable disease in the banana called Panama disease. There was no way to get the fungus out of the soil, so the Gros Michel was virtually wiped out across the planet by the 1950s.

Luckily, the Cavendish variety of banana was resistant to this fungus. British botanists found the Cavendish in southern China in the 19th century.

Now, as far as nature is concerned, we are all food for the Mill of Life. Two fungi have evolved and are now attacking the banana plantations of the Cavendish variety, as well as most other edible bananas.

One fungus, called Black Sigatoka, is threatening the Cavendish. The fungus first appeared in Fiji in 1963 and is spreading around the planet. It affects the leaves, and so the fruit ripens too soon. It reduces crops to one-third of the normal yield, and the number of productive years of the plant from 30 to just 2 or 3. Black Sigatoka can be killed with fungicides, but these chemicals are very nasty indeed to humans.

The other fungus, called Race 4, attacks the underground roots, which means fungicides don't work on it because they can't be sprayed on roots. This fungus, too, is spreading around the world.

Botanists are trying to evolve new types of banana that will be resistant to these fungi — but at the moment the new bananas don't have the 'right' taste and texture and are finicky about where they will grow.

Sexless Bananas

The original bananas with lots of seeds and hardly any flesh were fertile, and their seeds would grow into another banana plant if you planted them in the ground. Those bananas were diploid — each cell had two copies of each chromosome (just as we humans do). That sterile mutant of 10 000 years ago was triploid — it had had three copies of each chromosome.

Your regular yellow eating bananas of today don't have sex, which mixes up the DNA and gives genetic diversity. This has advantages and disadvantages. One advantage is that all the bananas of the next generation are the same as the previous generation, so they will all grow well — as long as the environment doesn't change. The disadvantage is that if something in the environment does change, and is unfavourable for the bananas, all the bananas will die. This is precisely what happened to the Gros Michel banana

It is hard, long and tedious work to 'evolve' new bananas, because bananas are sterile. Botanists have to find those rare mutant bananas that are not sterile. For example, in one search in Honduras, 30 000 plants had to be carefully examined to find just 15 fertile seeds.

The 'Other' Banana

The 'other' banana that is not often seen in the wealthy countries is the plantain. It has more starch than the 'regular' banana, and is often called 'cooking banana' or 'potato of the air', because it has to be cooked to make the starch edible.

References

'They Don't Grow On Trees', *Readers Digest Book of Fact*, Australia, 1994, p 286.

Pearce, Fred, 'Going Bananas', *New Scientist*, 18 January 2003, pp 26–29.

Encyclopaedia Britannica, Ultimate Reference Suite DVD, 2005.

Oranges and Vitamin C – the Real Juice

In high school, my son was in the rowing team. Part of his training was to try to bulk up some muscle, so he started pumping iron.

I warned him about some of the pitfalls. One problem is that you can bulk up muscle faster than you can bulk up the tendons that join the muscle to the bone, so the suddenly bigger muscles can tear and damage the tendons. So I advised him to eat lots of vitamin C to help the tendons grow more rapidly, and also said that getting it from food was probably better than getting it from a packet. He said, 'So I should eat oranges, the richest source of vitamin C', and I said, 'That's a mythconception'. Oranges do have vitamin C, but there are richer sources around.

Vitamin C

Vitamin C deficiency is one of the oldest nutritional diseases we know of; it used to be known as 'scurvy', and its symptoms

Capsicums vs Oranges

Oranges are rich in vitamin C, but capsicums are richer.

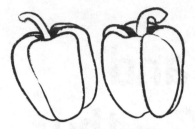

The Capsicum   **The Orange**

Vitamin C is essential in many human metabolic processes.
It's essential in the manufacture of collagen, which is
needed to make tendons, bones and healthy skin. It's
also used to make and release hormones to adrenal glands,
to maintain structural integrity of blood vessels and
much more.

include bleeding gums, loose teeth, generalised bleeding under the skin, slow wound healing and sore and stiff joints and legs. Scurvy was tragically far too common in sailors, who would spend long periods at sea without access to fruit or vegetables.

Vitamin C is essential for many human metabolic processes such as in the manufacture of collagen, which is needed to make tendons, bones and healthy skin (that's why the lack of it caused scurvy). It's also used to make and release hormones in the adrenal glands, to maintain the structural integrity of blood vessels, and much more.

The current recommended daily allowance (RDA) for vitamin C is about 50–100 mg, depending on which country you live in. Some scientists argue that the RDA should be higher, say, 200 mg.

Vitamin C was first isolated in 1928, and in 1932 was identified as the chemical that prevented and cured scurvy. In 1934, Sir Norman Haworth and Sir Edmund Hirst prepared vitamin C, making it the first vitamin to be artificially produced.

Megadose Therapy

For some, the 1960s and 1970s were a good time to turn back to Nature. Linus Pauling, a two-time Nobel Prize winner, wrote a series of books in the 1970s and 1980s in which he promoted megadoses of vitamin C as a cure for the common cold and many other diseases. Pauling was very convincing — having the authority that comes with winning two Nobel Prizes didn't hurt.

At the time I was very easily convinced of his theories (and somehow thought that vitamin C in powder form was very natural). I believed deeply in vitamin C and bought it in 25 kg lots; I was taking 10 000 mg (10 grams) per day (yup, that was my megadose).

The Reality

Today, the evidence seems to show that, while vitamin C doesn't reduce how many colds you get, it might reduce their duration by about half a day. It has other benefits too. A dietary survey of 23 000 people showed that if they didn't eat enought fruit and vegetables, they had twice the chance of developing inflammatory polyarthritis as compared to people who ate lots of fruit and vegies.

As a suitably vague statement, it's probably fair to say that a balanced diet will give you the vitamins you need. That's why today I don't worry about megadoses any more, and I am happy to get my vitamin C from food. The publication *Nutritional Values of Australian Foods*, written by Ruth English and Janine Lewis and published by the Australian Government Publishing Service, analyses Australian foods in great detail.

It says that for every 100 g of 'edible portion', oranges, lemons and limes all give you about the same amount of vitamin C — around 50 mg. Pawpaw and kiwi fruit come in at around 60 mg and 70 mg; parsley and chilli give you about 100 mg, but it's really hard to eat 100 g of parsley or chilli.

However, the real king of the vitamin C heap is the capsicum. Raw green capsicum will give you 90 mg, while raw red capsicum delivers 170 mg. So now I buy half a dozen capsicums in my weekly fruit shopping.

So, oranges are rich in vitamin C, but capsicums are richer. So the next time you feel like a vitamin C hit ... eat some capsicums.

How Little We Know ...

It took until 1753 before James Lind, a Scottish naval surgeon, showed that he could cure and prevent scurvy by feeding sailors citrus fruit such as oranges, lemons and limes. In fact, it was the limes that gave birth to the slang name of 'limey' for a British sailor.

Today, we know that vitamin C gets into the body and specific organs through receptors on cells. The Sodium-Coupled Vitamin C Transporters Type 1 (SVCT 1) are involved in bulk transport around the body, while the Sodium-Coupled Vitamin C Transporters Type 2 (SVCT 2) help vitamin C get into specific organs.

The SVCT 1 receptors help maintain the levels of vitamin C in the blood plasma between 10 and 160 mM (millimoles). But in other organs such as the adrenal glands, lungs, pancreas, spleen, testes, ovaries and eyes, the vitamin C levels are at least 100 times higher. This seems to be because of the SVCT 2 receptors.

References
English, Ruth and Lewis, Janine, *Nutritional Values of Australian Foods*, Australian Government Publishing Service, Canberra, 1991, ISBN 0644138718.

Hediger, Matthias A., 'New View at C', *Nature Medicine*, May 2002, pp 445, 446.

Pattison, Dr D.J. et al., 'Vitamin C and the risk of developing inflammatory polyarthritis prospective nested case-control study', *Annals of Rheumatic Diseases*, July 2004, pp 843–847.

Chewing Gum:
the Sticking Point

We humans have been chewing gum for thousands of years with very few side effects. Even so, many of us believe that if we swallow chewing gum it will sit, undigested, in the gut for seven years. In some cases, it might even block the gut. However, the first time I heard this my suspicions were immediately raised by the use of that nice prime number, seven. You see, 'seven' pops up in all kinds of psycho-babble pseudo-science, including some of the motivational books and programs that offer you Seven Steps to a New and Improved You.

History of Chewing Gum

Archaeologists have found 9000-year-old lumps of black tar, and the bite impressions on them suggest that most of the chewers were kids aged 6 to 15. Two thousand years ago the Greeks chewed the pale yellow resin from the mastic tree, while American Indians chewed spruce gum.

Chewing gum came to modern America via the Mexican general Antonio Lopez de Santa Anna, who was responsible for the massacre at the Alamo in San Antonio, Texas, in March 1836. Santa Anna later entered the USA and settled on Staten Island, New York. He brought with him a big lump of chicle, the dried milky sap or latex of the Mexican sapodilla tree, which the inhabitants of what is now called Mexico had chewed for thousands of years.

A local New York inventor, Thomas Adams, tried unsuccessfully to turn the chicle into a cheap rubber — but then he remembered how his son and Santa Anna loved to chew chicle together. In February 1871 small, tasteless balls of chicle were first sold in New Jersey as 'Adams New York Gum — Snapping and Stretching'.

Side Effects (Good and Bad)

One of the nice side effects of chewing gum is that you increase the production of saliva in the mouth, which is usually good for oral hygiene and your breath. This saliva can then neutralise some of the acids in certain foods.

The bad (and relatively uncommon) side effects of chewing gum include diarrhoea, tummy pain and flatulence (from the sorbitol in sugarless gum). Even more uncommon are mouth ulcers (from cinnamon flavouring), high blood pressure and low blood potassium (from liquorice flavouring).

A very rare side effect is higher blood mercury levels — from the dental amalgam in fillings already in your mouth. The mercury leaches out of the amalgam into the saliva, and then into the bloodstream. But this would happen only in cases of excessive chewing. A study of 18 chewers who chewed a median of ten nicotine chewing gum sticks per day showed a jump in plasma mercury levels from 4.9 nmol/L to 27 nmol/L.

Other unpleasant side effects include mechanical injury to the teeth, overuse injury (including temporomandibular joint syndrome) and even the extrusion of dental repairs. But the overwhelming majority of us happily chew gum with no harmful side effects whatsoever.

You are what you chew

What does chewing gum consist of?

Chewing Gum

Flavours:
Liquorice
Menthol
Eucalyptus
Peppermint
Freeze-dried fruit
Strawberry
Lemon
Apple
Blackcurrant
Cola
Orange

Sweeteners and
active ingredients:
Sorbitol (extracted from corn
or wheat)
Maltitol (extracted from corn
or birch bark)
Isomalt (extracted from sugar)
Aspartame (chemically produced
sweetening effect 200 times
higher than sugar)
Sugar
Propolis
Baking soda
Dextrose
Calcium
Zinc
Vitamins
Fluoride

Mind you, chewing gum in public places has given us the word 'gumfitti', meaning 'gum pollution in public places'. Down on the chemical level, chewing-gum base is made from long chains of molecules, and these chains have very few locations that will allow oxygen in, so they are resistant to oxygen, and also to ultraviolet light. So chewing-gum base is pretty high on the non-biodegradable scale. Thus gumfitti has created a whole new expensive industry involving fancy gum-removal devices (solvents, liquid nitrogen, etc). Some councils in Ireland spend up to €54 000 each year to remove chewing gum from public places. Singapore has even made most non-medical uses of chewing gum illegal.

What is Gum?

Modern chewing gum has five basic ingredients. First is the gum base — the chewy bit. It used to be totally natural, but today is usually a mix of natural (chicle) and synthetic (such as polyvinyl acetate) gums. The other four ingredients are softeners (such as vegetable oils), flavours, sweeteners and corn syrup. The saliva in your mouth will dissolve all of these, except for the gum.

Gum is excellent at sticking to hair and shoes — but does gum really stick to the wet and slippery lining of your gut? Nope, it comes out with the rest of your solid wastes — through the same pathway and, almost always, right on time.

Gum Blocks Gut

However, we humans are a very varied bunch. You have to expect that there might be some people, somewhere on the planet, in whom chewing gum might get trapped. After an exhaustive search, I tracked down the paper 'Chewing Gum Bezoars of the Gastrointestinal Tract', by Dr David Milov and his colleagues, in the journal *Paediatrics*.

Dr Milov discussed three of his patients (aged 18 months to four-and-a-half years) who had developed obstruction of the gut specifically because they swallowed gum. But these very rare cases were not kids who had just accidentally swallowed some gum once in their life.

For example, the 18-month old (who had a long history of chewing gum, despite her young age) had also swallowed four coins. The other two children had a history of swallowing up to seven pieces of gum per day — they did this so that somebody would give them another piece. One of the kids had been constipated for two years ... so maybe that's how long the chewing gum had been clogging up his gut.

Dr Milov noted that the 'rectal masses' had to be manually removed, and that they displayed the characteristic 'taffy-pull sign' — a long, multicoloured, skinny trail of gum — as the doctor pulled out a small lump of the mass.

Gum is Mostly Safe

There are, however, only a handful of such cases in the medical literature, and only in young kids. There was absolutely no evidence to support the myth of gum's seven-year residency in the gut of the average person. And when obstruction happened in kids, it was usually in kids who had a previous history of constipation or poor digestion that existed long before they started swallowing huge quantities of chewing gum.

The rest of us can walk and chew gum at the same time, and we can occasionally swallow it with no problem at all.

TMJ Syndrome

Temporomandibular joint (TMJ) syndrome refers to the joint at the side of your face, just below your ear, where your jaw is joined to your skull. Each time you open your mouth (to eat, speak, chew gum, yawn, etc.), you move this joint. Like any moving part, it can wear out.

TMJ is a complicated condition, with many symptoms, ranging from very mild to very nasty. You can have chronic headaches, neck aches, earaches, pain in the shoulders and back, ringing in the ears, dizziness, stiff or locking jaws, or clicking, popping or grinding sensations in the joint.

It seems that the traditional father of medicine, the Greek physician Hippocrates (460–377 BC), knew of TMJ syndrome. He described it happening in 'a group of patients whose teeth are disposed irregularly, crowding one on the other, and they are molested by headaches.'

Very Obscure Side Effect

After I wrote a short story about chewing gum, I received this email from a very satisfied reader. It said:

* * *

Hi Karl — Thank you and thank you again — you have solved a medical/dental mystery for me. In May and November/December last year for the first time in my life I had trouble with mouth ulcers and gum infections — leading to considerable mouth pain, several visits to doctor and dentist and consumption of courses of Amoxil. After examinations and X-rays, no cause could be found. Then this morning I read your column in The Age *weekend mag. about gum — in which you mention that cinnamon gum can cause mouth ulcers — the mystery has been retrospectively solved. Now, I can connect these outbreaks to the cinnamon gum we bought in Alaska last May. I had the first ulcers just after we bought the stuff. By the time we returned to Australia it had cleared up — and then in November my daughter found a few packs of leftover gum, which she didn't like and which I then occasionally sampled. Suddenly I had mouth ulcers and an ensuing infection again. Since the gum has gone so have the ulcers.*

* * *

First, this man was a professional wine taster, so having ulcers in the mouth would have been disastrous for his career. Second, this is a good example of how tricky a job your general practitioner doctor can have.

References

Allen, C.M., et al. 'Oral mucosal reactions to cinnamon-flavoured chewing gum', *Journal of the American Dental Association*, May 1988, pp 664–667.

Lindley, Geraldine, 'Was it something you ate?', *British Medical Journal*, 11 January 2003, p 87.

Milov, David E., et al., 'Chewing Gum Bezoars of the Gastrointestinal Tract', *Paediatrics*, August 1998.

Sallsten, G., et al., 'Long-term use of nicotine chewing gun and mercury exposure from dental amalgam fillings', *Journal of Dental Research*, vol. 75, 1996, pp 594–598.

Fit to Burst: Do Bodies Explode in a Vacuum?

The Governor of California, Arnold Schwarzenegger, had previous careers in body building and making movies. In his movie *Total Recall*, the hero, Quade (played by Arnie), begins to physically expand when he is exposed to the incredibly thin atmosphere of Mars. In a desperate effort to survive, he holds his breath — and instantly his face bulges horribly and his eyeballs begin to swell dangerously, as he heads down the pathway towards exploding. Luckily, an ancient alien artefact chugs into activity right on cue and almost instantaneously manufactures enough atmosphere to restore him to normal size and let him breathe again.

So here's the question: will we humans explode in the full vacuum of space, as urban legends claim? The answer is no. We won't explode and, if the exposure is short enough, we can even survive.

So why do people think that we explode when exposed to a vacuum?

The bulge vs the explosion

The question is ... will humans explode in the full vacuum of space? We are lead to believe that they will POP.

The Hollywood way

What actually happens
(there will be some bulging though)

'Explosive' Decompression

Maybe the idea started from the term 'explosive decompression', which is used to describe a plane suddenly losing pressure inside.

In the aerospace industry, there are three types of decompression. 'Explosive decompression' happens when all the air in a smallish plane is lost in less than half a second. This is usually accompanied by a loud noise (technically known as a 'bang'). More common is the 'rapid decompression' that can happen in larger planes. This takes longer, between half a second and ten seconds. Finally, there is the 'subtle decompression' that takes longer than ten seconds. This can be caused by the cabin altitude pressure being set wrongly, or the valves freezing in the 'open' position (see passage within 'Jetsetting Germs' on page 10). The danger in these last two types of decompression is that you might not notice it happening, and could fall unconscious.

So maybe the choice of the word 'explosive' in the phrase 'explosive decompression' is responsible for the mythconception that people would also go bang (or explode) in a vacuum. But, in truth, the term 'explosive decompression' specifically refers to how suddenly the pressure drops, not what happens to the people exposed to that drop in pressure.

20 Tonnes of Pressure

In addition, there is the totally correct belief that the pressure inside our bodies pushes against the weight of the air pressing against our skin, so putting us in a state of perfect balance.

At ground level, the air pressure is about ten tonnes per square metre. One way of looking at this is that the air pushes with a 'weight' of 10 tonnes on each square metre. If we measure out a 1 m² in the back yard and weigh all the air above that square — going right up to space — we will find that all the air in that column weighs about 10 tonnes. So to create a balance against this weight of air, each square metre of our body pushes back with 10 tonnes. A tall person might have a surface area of nearly 2 m², which means that the total weight of air on their skin is about 20 tonnes.

So if we remove the external pressure, would the internal pressure of our body's tissues make our body explode?

What, Me Explode?

The answer is no. Skin is quite tough (after all, they make leather out of animal skin), and it will not split.

The body can expand to nearly twice its usual volume, but it will not burst asunder. It expands because the water in the tissues will rapidly turn into water vapour. The air in the gut (about 1 l in volume) will expand and cause tummy pain. It will also push the diaphragm upwards, making it harder to breathe and putting stress on the heart. The air will rush out of your lungs via your mouth, probably damaging some of the delicate, feathery lung tissue. If you foolishly try to hold your breath (as Quade did), bubbles of air will force their way into your bloodstream, and then to your brain, causing a stroke.

Air can be trapped inside your middle ear, and your sinuses, causing pain. You will also suffer decompression sickness (called 'the bends' due to bubbles of nitrogen in the muscles and bones which make you bend over to relieve the pain) and lack of oxygen (called 'hypoxia').

By the way, the air pressure on Mt Everest (28 000 ft (8.5 km) above sea level) is about one-third of sea level air pressure; most untrained people exposed suddenly to that pressure will fall unconscious within three to six minutes thanks to hypoxia.

In one case in 1982, a technician testing a vacuum chamber was accidentally exposed to pressure equivalent to an altitude of over 74 000 ft (about 22.5 km) for over one minute. Now, the air pressure at 74 000 ft is about 3.6% of the pressure at sea level — in other words, close to zero. When he was pulled out he was blue in colour, frothing at the lips and bleeding from the lungs — but he hadn't exploded. He also recovered fully. But there are also cases of people who have been exposed for longer periods, and who have died.

2001 Was Right

The scenario in the 1968 movie *2001: A Space Odyssey* was surprisingly accurate. The astronaut David Bowman is locked out of his spaceship by the crazy computer HAL 9000. He bursts into his spaceship by leaping through the vacuum of space from his spacepod, and is exposed to that vacuum for about ten seconds. It turns out that in a vacuum you will stay conscious for about 10–15 seconds, and you can survive a 90-second exposure. But be sure not to hold your breath.

You will not explode, but you may expand — even more than a fully pumped-up Arnold Schwarzenegger could ever dream of.

HAL

In *2001: A Space Odyssey*, the complicated spaceship is run by the intelligent computer HAL 9000. At one stage HAL introduces himself to the human crew with the words, 'Good afternoon, gentlemen. I am a HAL 9000 computer. I became operational at the HAL lab in Urbana, Illinois, on the 12th of January, 1992.' 'HAL' stands for Heuristic Algorithmic, which represent the two main processes of learning.

But what about the rumour that HAL was chosen as a name because each letter of that acronym is one letter less than the name of the then-dominant computer company, IBM? Totally wrong. Arthur C. Clarke (the science-fiction writer who co-wrote the movie with Stanley Kubrick, the movie's director) said that neither he nor Stanley Kubrick were aware of this coincidence at the time.

Expanding Hand

In 1960, Joe Kittinger Jr carried out the highest-ever parachute jump. He ascended to some 103 000 ft (almost 32 km) in an unpressurised balloon while wearing a primitive spacesuit. His right glove sprang a leak. He later wrote in *National Geographic* magazine, 'I examine the pressure glove; its air bladder is not inflating ... From my previous experiences, I know that the hand will swell, lose most of its circulation, and cause extreme pain ... I decide to continue the ascent, without notifying ground control of my difficulty.' He definitely did not want to tell ground control, because it took a huge amount of effort and organisation to get him on this balloon on that day. If he'd cancelled, there would not be another flight for a long time and, almost certainly, he would not be on it.

Kittinger eventually reached 102 800 ft. He wrote, 'Circulation has almost stopped in my unpressurized [sic] right hand, which feels stiff and painful.'

He jumped. In that 13-minute-and-45-second jump, he set three records.

First, he set the record for the highest parachute jump — 31.3 km. Second, he set the record for the longest parachute free fall — four minutes and 36 seconds (the time before he opened the parachute). Third, he became the first person to exceed the speed of sound without using an aircraft or a space vehicle. In the thin upper atmosphere, he reached a speed of some 1149 kph.

And he did it all with a right hand the size of a boxing glove. Three hours after he landed his swollen hand went back down, with no ill effect.

References

Adams, Cecil, 'If you were thrown into the vacuum of space with no space suit, would you explode?', 'The Straight Dope'. http://www.straightdope.com/classics/a3_147.html

Kittinger, Joseph W. Jr, 'The Long, Lonely Leap', *National Geographic*, December 1960, pp 854–873.

Shayler, David J., *Disasters and Accidents in Manned Spaceflight*, Springer-Praxis Books in Astronomy and Space Science, Chichester, UK, 2000.

The Amityville Horror Story – Boo!

If you wanted to go to the movies and get scared by the latest horror movie, the 1970s and '80s were perfect. The white-knuckle brigade scared themselves silly with movies such as *Rosemary's Baby*, the *Omen* series, *Exorcist I* and *II*, the *Friday the 13th* series, all the Hammer horror movies — and, of course, the Big One: *The Amityville Horror*. There are lots of scary films around, but the special thing about *The Amityville Horror* is the claim that it was based on fact. Unfortunately, this claim has turned out to be false.

The original Amityville movie (released in 1979) wasn't the best horror movie ever made, but it grossed over US$80 million — at the time this was the best ever result for a film from an independent studio, a record the film would hold for a decade. Part of the reason for its success was the widely believed (and widely promoted) claim that the movie told a true story.

Amityville House History

The Amityville story began in 1924, when John and Catherine Moynahan built the house at 112 Ocean Avenue, Amityville, on

Long Island in New York. They lived there happily with their children for many years. In 1965 their descendants sold the house to the DeFeo family, who renovated it, landscaped it and put in a swimming pool. Apparently the DeFeo family was moderately dysfunctional, and often used violence to resolve situations. The father, Ronald, was brutal; Ronald Jr, the eldest son — otherwise known as 'Butch' — was soon using drugs and alcohol, developed a heroin addiction and was always getting into trouble with the local police.

Nine years after the DeFeo family moved in, on 13 November 1974, Ronald 'Butch' DeFeo Jr rushed into Henry's Bar in Amityville at 6.30 p.m., and yelled out, 'You've got to help me! I think my mother and father are shot.' Five of the patrons piled into Ronald's car for the short and scarily speedy drive to the house. There they found his parents, Ronald Sr and Louise, and his two brothers and two sisters, all shot dead. It transpired that Ronald had long hated his father, and had even planned how to kill him. The son was found guilty of the murders and sent to prison for six consecutive sentences of 25 years to life — a minimum of 150 years.

The Lutz Family

On 18 December 1975, two weeks after Ronald had been sentenced, the Lutz family — newlyweds George and Kathleen (Kathy) Lutz, and her three children — moved into 112 Ocean Avenue. The house would ordinarily have been too expensive for them — for one thing, this huge Dutch Colonial six-bedroom mansion was on the water. But, on the other hand, it was rather creepy-looking in a Gothic kind of way, and it did have the terrible history of the DeFeo murders. So it sold for the amazingly low price of US$80 000.

Within a month they had abandoned the house, leaving behind all their possessions, and moved in with Mrs Lutz's mother.

The Lutzes claimed that they had been terrified by demonic manifestations, obviously related to the monstrous murders.

Cue the 'demonic manifestation'

The 3-step recipe for some scary success.

1

2

3

Take one kinda scary-looking house with some just plain weird 'eye' windows.

Add a well-known demonologist, Ed (shown) and his wife Lorraine Warren (not shown).

Stir well with plenty of wine.

... and shazzam – instant Hollywood scary story.

These included levitations, strange sounds, floating pigs (with blood-red eyes) that left footprints in the snow outside and green slime dripping down walls; a priest suffered blisters on his hands whenever he tried to enter the house, and there was a hidden, satanic, red-painted underground room that smelt of blood, doors torn off hinges, etc. (The floating red-eyed pig was called Jodie, and could change its size from that of a teddy bear to bigger than a house.)

The Book, The Movie

In September 1977, Jay Anson released his book, *The Amityville Horror: A True Story*. The publisher, Prentice Hall, claimed it was 'the non-fiction *Exorcist*'. The book was based on 35 hours of taped commentary from the Lutz family, and it sold three million copies over some 13 printings. The copyright page read, 'All facts and events, as far as we have been able to verify them, are strictly accurate.' However, Jay Anson later admitted that he did not verify any of the stories — he simply compiled the book from the tapes he was given.

The first Amityville movie was released in 1979. It was followed by some dodgy sequels — *Amityville II: The Possession* and *Amityville 3-D* (also called *The Demon*) — and five straight-to-video movies.

Qui Bono? (= Who Benefits?)

Practically everybody did well out of the book and movies (except the dead family, and Ronald DeFeo Jr, who was in prison) as this story became probably the most famous demonic possession case of all time.

Jay Anson grew wealthy from selling three million books, and from the movie rights.

The parapsychologists who had been called in, Ed and Lorraine Warren, got lots of prestige and publicity.

The Lutz family enjoyed both the fame that came from being demonically haunted and the thousands of dollars that came from book and movie contracts. As it turns out, the money was very handy indeed, because they had not been able to make one single mortgage payment on this remarkably cheap house. Could this be the *real* reason they left their house?

William Weber, Ronald DeFeo Jr's lawyer, benefited twice. First, he wanted the chance to use an insanity plea to argue for a retrial for his client, using the excuse that evil spirits had driven Ronald to kill his family. The judge threw the excuse out of court, but at least Weber tried. Second, the Lutz family had double-crossed Weber and booted him out of the book and movie deals, in which he had been originally involved. So he teamed up with Hans Hotzer to co-author the book *Murder in Amityville*, which became the basis of the second movie, *Amityville II: The Possession*.

The Real Story

The real story was told by Dr Stephen Kaplan in his fully referenced book, *The Amityville Conspiracy*. Unlike many others in the field of parapsychology, he deserved the title of 'Doctor', as he held a PhD in sociology from Pacific College.

As a parapsychologist, Dr Kaplan had been invited by the Lutz family to investigate the happenings at their house very early in the proceedings. But when he pointed out that if he found a hoax, he would expose it, the Lutz family got rid of him and quickly replaced him with the well-known demonologists Ed and Lorraine Warren. Steve Kaplan was vindicated when DeFeo's lawyer, William Weber, admitted later in sworn testimony that he and the Lutzes had concocted the story of the demonic house over several bottles of wine. He told Associated Press, 'We created this horror story over many bottles of wine that George Lutz was drinking. We were creating something that the public wanted to hear.' In *People* magazine of 17 September 1979, Weber said of the Jay Anson book, 'I know this book's a hoax. We created this horror story over many bottles of wine.'

But surely the Lutz family had some independent proof, or eyewitness reports? The answer is no. George and Kathleen Lutz were the only witnesses to all the demonic happenings. According to police records, the Lutz family did not call the police for help at any time during the three to four weeks of their supposed hauntings. However, the movies and the books do claim that the police were called in.

Total Fabrication

In fact, every single 'demonic manifestation' in the house that could be independently tested or verified was disproved.

The doors and windows that had supposedly been torn off their hinges or damaged showed absolutely no sign of damage. The demonic pig's footprints that supposedly appeared in the snow happened when the weather records showed there had been no snowfall. The third-floor window in the shape of a quarter moon that had supposedly been cracked showed no signs of cracks or repairs. The demonic 'red room' that was supposed to be some kind of sinister dungeon or hideaway was merely a plumbing access space. The priest denied having blisters appear on his hands. And so on.

The Lutz family later admitted that the whole story of the hauntings was a fabrication during the following, very messy, trials. One trial had their ex-lawyer, Weber, suing them for squeezing him out of the book deal — a $2 million lawsuit. Another trial came about because later owners of the house sued the Lutzes, Jay Anson and his book publishers for fraudulent claims of hauntings that led directly to their later loss of privacy in their house. At all hours of the day or night, people would trample across their yard — and they were sick of it.

The Dust Settles

The murderer, Ronald DeFeo Jr, was later asked what he thought of the whole series of Amityville books, movies and videos. From prison he said, 'The only thing that's real were the murders ... yes, it's all a hoax ... it's all about money ... a cold-blooded murder, period ... no ghosts, no demons ... This ain't funny no more, people look in my eyes like I'm possessed or something, I'm sick of it.' He also wrote in a letter, 'Amityville was a hoax that Weber and the Lutzes started. Yes, to make money. It started as my trial was in progress.'

Also, various people had been falsely credited as witnesses to the demonic happenings, and they didn't like this. So they threatened the Lutz family and Jay Anson with court actions. As a result, they were quietly written out in later editions of the book. These people included Marvin Scott (an anchorman for the local Channel 5 TV News), Steve Bauman (reporter) and Sergeant Pat Cammaroto (Amityville Police Department).

The books make false claims about the site of the house being used as an asylum by the Indians, before Europeans moved in. None of the people who lived in the house before the Lutzes noticed any demonic happenings. Immediately after the Lutz family moved out, Barbara and James Cromarty bought the mortgage and moved in. Neither they nor any of the families who later lived in the house had any problems with demons. The only troubles they had were with curious tourists, gawkers and rubberneckers, so later owners renumbered the house and changed the rather distinctive windows.

It's Fun to Be Scared …

The 'true story' claim was repeated for the 2005 movie remake starring ex-*Home and Away* star Melissa George. Maybe that claim helped the movie go straight to the top five of the US Box Office list in its first week, during which time it dragged in over US$30 milllion.

And maybe another reason for its success is that some of us really like to be scared — but only in a safe way, at the movies.

Recycled Horror Music

A few soundtracks were tried out for *The Exorcist*. One of the rejected soundtracks was recycled for the latest *Amityville Horror*. That factoid should win some points at a trivia night …

References

Anson, Jay, *The Amityville Horror*, Prentice Hall, 1977, ISBN 0130325996.

Holzer, Hans, *Murder in Amityville*, Belmont Tower Books, 1979, ISBN 0505514087.

Kaplan, Stephen and Salch Kaplan, Roxanne, *The Amityville Horror Conspiracy*, Toad Hall Inc, 1995, ISBN 0963749803.

Nickell, Joe, 'Investigative Files Amityville: The Horror of It All', *Skeptical Inquirer* January/February 2003.

Folding Paper – the Plane Truth

When my son was near the end of his primary school years, I thought it was time to pass on some of my Weird Freaky Science Wisdom, and have a little bit of fun as well. I told him that I would give him a million dollars if he could fold a piece of paper in half, and in half again, and so on for a total of ten times. Of course he tried, and of course he failed.

I knew that this would happen, because it was Accepted Wisdom that it was impossible to fold a piece of paper in half ten times (or seven, or nine, for that matter). I told him that it couldn't be done, even if he used paper the size of a football field. But I now know that I was wrong.

Start Folding

Suppose you start with an average A4 sheet of paper about 300 mm long and 0.05 mm thick.

The first time you fold it in half it becomes 150 mm long and 0.1 mm thick. The second fold takes it to 75 mm long and 0.2 mm thick. By the eighth fold (if you can get there), you have a blob of paper 1.25 mm long and 12.8 mm thick. It's now thicker than it is

7 <u>was</u> the magic number

It is accepted wisdom that it is impossible to fold a piece of paper in half more than seven times.

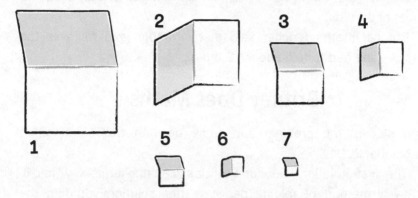

Suppose you start with an A4 piece of paper around 300 mm long and 0.05 mm thick. The first fold makes the paper 150 mm long and 0.1 mm thick. By the eigth fold (if you can get there), you have a knuckle of paper 1.25 mm long and 12.8 mm thick. It's now thicker than it is long and practically impossible to bend.

long and, if you're trying to bend it, seems to have the structural integrity of steel.

A typical claim on the Internet might run, 'No matter its size or thickness, no piece of paper can be folded in half more than seven times', and, as you stare sadly at your block of folded paper, you might tend to agree.

Britney Does Gold

And so Accepted Wisdom ruled — until 2001.

That was when a high-school student, Britney Gallivan of Pomona, California, was given a maths problem. She was told she'd get an extra maths credit if she took up the option of solving the problem of folding a sheet of paper in half 12 times. She tried and failed with average-sized sheets of paper.

So she got smart and used something incredibly thin: gold foil, only 0.00028 mm thick. She started with a square sheet, 10 cm by 10 cm. It took lots of determination and practice, as well as rulers, soft paintbrushes and tweezers, but Britney finally succeeded in folding her gold foil in half 12 times. She ended up with a square sheet of gold foil.

But her maths teacher said that ultra-thin gold foil was too easy — she had to fold *paper* 12 times.

Britney Does Maths

She studied the problem and came up with two solutions in December 2001.

The first solution was for the classical fold-it-this-way, fold-it-that-way method of folding paper. In her solution you fold the paper in alternate directions. She derived this formula:

$$W = \pi\, t\, 2^{3(n-1)/2}$$

where 'W' is the width of the square sheet that you start with, 't' is the material's thickness and 'n' is the number of folds possible.

The second solution was for folding the paper in a single direction. This is the case when you try to fold a long, narrow sheet of paper. Britney derived a sheet-folding formula:

$$L = \frac{\pi \cdot t}{6} \cdot (2^n + 4)(2^n - 1)$$

where 'L' is the minimum possible length of material, 't' is the material's thickness and 'n' is the number of folds possible in one direction.

When Britney looked closely, she found that if you are trying to fold the sheet as many times as possible, there are advantages in using a long, narrow sheet of paper.

Britney Does Paper

The formula told her that to successfully fold paper 12 times, she would need about 1.2 km of paper.

After some searching, she found a roll of special toilet paper that would suit her needs — and which cost US$85. In January

2002 she went to the local shopping mall in Pomona. Along with her parents, she rolled out the jumbo toilet paper, marked the halfway point and folded it the first time. It took a while, too, because it was a long way to the end of the paper …

Britney folded the paper the second time, and then again and again.

After seven hours, she folded her paper for the 11th time into a skinny slab about 80 cm wide and 40 cm high, and posed for photos. She then folded it another time (to get that 12th fold essential for her extra maths credit), and wrote up her achievement for the Historical Society of Pomona in her 40-page pamphlet, *How to Fold Paper in Half Twelve Times: An 'Impossible Challenge' Solved and Explained*. She wrote in her pamphlet, 'The world was a great place when I made the twelfth fold.'

Britney succeeded because she was as contrary and determined as Juan Ramon Jiminez, the Spanish poet and winner of the 1956 Nobel Prize for Literature, who wrote, 'If they give you ruled paper, write the other way'.

Fold 50 Times?

Here's a great dinner party question: if you had a sheet of paper and folded it in half 50 times, how thick would it be?

The answer is about 100 million kilometres, which is two-thirds of the distance between Earth and the Sun — and nobody ever guesses anything close to this.

References

Gallivan, B.C., *How to Fold Paper in Half Twelve Times: An 'Impossible Challenge' Solved and Explained*, Historical Society of Pomona Valley 2002.

Peterson, Ivars, 'Folding Paper in Half — Twelve Times', *Science News*, 24 January 2004.

Peterson, Ivars, 'Champion Paper-Folder', *Muse*, July/August 2004, p 33.

William Tell Tale

Most people have heard of the legend of the famous Swiss patriot William Tell. The story runs that in AD 1307 William Tell, a farmer and hunter, was asked to take off his hat as a mark of respect to the foreign Hapsburgs who were then running the country. He refused, and was then forced to shoot a crossbow bolt at an apple perched on his son's head.

William Tell split the apple, was arrested anyhow, bound, taken in a boat across Lake Lucerne to be incarcerated in a dungeon, released from his bonds to guide the boat to safety from a sudden storm that had sprung up, leapt to safety, kicked the boat back into the waves, escaped, and then helped to set up the federation of cantons that is modern Switzerland. It's a great tale, but it's all a fable.

The Really Old Days

People migrated into what is now Switzerland several thousand years ago. Once they were stopped by the icy mountains and narrow Alpine valleys, they settled down in the valleys. The sheer mountains made very good barriers between one valley and the next. But the climate began to warm around 1000 AD, creating more pastureland, which made the countryside more habitable, and the population correspondingly increased.

The closest trading partner was Italy, just over the Alps. The standard trade route was to the south, through the St Gotthard Pass. The route north was blocked by an impassable gorge, but in the mid-13th century, the men of the canton of Uri made a bridge across this gorge. Now there was a trade route between northern and southern Europe, with the St Gotthard Pass and the new bridge as critical waypoints. Suddenly the people of Uri had travellers spending three days passing through their lands, and spending money on accommodation and supplies on the way through.

Uri became wealthier, but also began to suffer internal battles between a few clans. The community of Uri asked a neighbour, Count Rudolph von Hapsburg, to be a mediator. The count came in with his soldiers, and wouldn't go away.

William, tell me who it is?

It appears that the story of William Tell may have sprung from the fertile imaginations of the creators of some earlier Danish fables.

Still, it's not a bad little story!

Tell — the Good Old Days

The story goes that in 1307, William Tell and his son walked into the central market square of Altdorf; Altdorf was (and still is) the capital of the canton of Uri. The Hapsburg Duke of Austria was represented in Altdorf by the Bailiff Gessler (equivalent to a governor). On this day, he placed a Hapsburg hat on a pole in the market square and demanded that everybody present had to show their respect for the Hapsburg rulers by taking off their hats.

William Tell refused.

Gessler was infuriated. To teach Tell a lesson, he had Tell dragged to one end of the square and Tell's son dragged to the other, 120 m away, where Gessler had an apple placed on the boy's head. If Tell could not shoot a single crossbow bolt into the apple, both Tell and his son would be killed. Of course if Tell aimed too low, the bolt would kill his own son.

Tell accurately split the apple, so Gessler told Tell that he and his son were free to go, but asked why Tell had slipped a second bolt into his shirt. Tell told Gessler that if he had missed, he would have killed Gessler with that second bolt. This led to Tell's being bound, taken across the lake, etc.

Tell — the Publicity Machine

The Swiss historian Aegidius (or Gilg) Tschudi was the first person to mention William Tell in 1569, which was some 262 years after the supposed incident. Tschudi wrote the *Chronicon Helveticum*, which covers the years 1000–1470 AD. However, the *Encyclopaedia Britannica* does not have a very high opinion of the *Chronicon Helveticum* as an accurate historical record. It writes, 'Many assiduously collected documents were incorporated into it; others were fabricated, in an attempt to give a coherent and comprehensive chronology'.

In the mid-1700s, Gottlieb de Haller, a scholar from Bern, read an ancient Danish folktale about King Harald Bluetooth (who reigned 936–987 AD) and a Viking chieftain called Toko. The story had Toko being forced to shoot an apple off the head of his little son. This was suspiciously similar to the William Tell story. De Haller then wrote *William Tell: A Danish Fable*, in which he claimed that the story of William Tell was based on this Toko tale. It even had the same details of the apple on the son's head and the extra arrow hidden in the father's shirt. But the locals liked their history, regardless of whether or not it was real, and de Haller had to backpedal furiously after a court action and personal threats, and his book was publicly burnt in the market square in Altdorf. He apologised deeply and often, saying that the book was not a serious work of history, just a playful literary exercise.

The great German dramatist Friedrich von Schiller certainly didn't care whether or not the Tell tale was real — he could see that it was a great story. He wrote a box office hit play about William Tell, which his friend Johann Wolfgang von Goethe directed when it first opened in March 1804. Rossini could also recognise a great story, so in 1829 he wrote the opera *William Tell.*

Tell Today

In modern times, movies and, later, television picked up the Tell theme. In 1940, Hollywood produced the animated film *Popeye Meets William Tell*. Even the Nazis drastically rewrote William Tell into a propaganda movie. In 1954, Errol Flynn bankrolled and starred in a film called *William Tell*, but it was never finished. The 'William Tell Overture' then became the theme for *The Lone Ranger* radio and TV series, and today's average American will hear the tune 573 times during their lifetime.

The public perception of William Tell is that he was real. The William Tell crossbow appears on every item exported from Switzerland as a mark of authenticity. A bronze statue has been erected to Tell's memory in the market square of Altdorf, to commemorate the original crossbow-apple incident.

But back in the 1700s, Gottlieb de Haller's book threw the light of scepticism on the sacred legend. Unfortunately for the legend, later historians discovered that they could find absolutely no documentation that William Tell had ever existed. The very latest 1000-page history of Switzerland gives just 20 lines to William Tell — although it does put a photo of a bronze statue of him on the cover.

The Truth

So let's completely bypass any controversy with a different slant from Dan Rather, the American TV newscaster, who said, 'An intellectual snob is someone who can listen to the "William Tell Overture" and not think of *The Lone Ranger*.'

Switzerland Arises

After William Tell supposedly escaped from his captors on the lake in the storm, he travelled some 30 km to the pass of Hohle Gasse, knowing that Gessler would have to come through it. He was correct, and he shot Gessler with the second bolt he had hidden in his shirt.

Tell later met with three other men, representing three neighbouring cantons, in a forest meadow today known and revered as Rütli. There they swore the sacred oath that led to the combining of the cantons.

The oath says: 'To assist each other with aid and every counsel and every favour, with person and goods, with might and main, against one and all, who may inflict on them any violence, molestation or injury, or may plot any evil against their persons or goods'. As a direct result of this meeting, sporadic and then concerted fighting arose against the foreign occupiers. Soon cantons grouped together, leading ultimately to the Switzerland of today.

Unfortunately for the Tell legend, a copy of the original Oath of Rütli was found in 1758. It bore the names of three men, not four.

Even worse, none of the men was named Tell, and there was no mention of that name anywhere on the document. Even worse for the reputation of the historian Tschudi, the document was dated 'the beginning of August 1291' — not 1307.

Even so, there is a statue of Tell and his son in the market square of Altdorf. It seems that we need our legends more than we need the truth.

References

Wernick, Robert, 'In Search of William Tell', *Smithsonian*, August 2004, pp 70–78.

Encyclopaedia Britannica, Ultimate Reference Suite DVD, 2005.

Water Down the Drain – Plughole Science

If you've ever looked at a weather map, you might have noticed that in the southern hemisphere all cyclones spin clockwise. According to popular lore, so does the water in toilet bowls, bathtubs and hand basins. It is true that a cyclone will spin clockwise in the southern hemisphere, but as for the water spinning only one way as it goes down a drain ... well, that's another myth (unless you do the experiment very, very carefully).

Coriolis Force (Hard Stuff, Okay to Skip)

The force that makes cyclones spin is called the Coriolis Force, named after Gustave-Gaspard Coriolis, who first described it way back in 1835. A Real Physicist will tell you that there's no such thing as Coriolis Force — it's all just a fancy name for 'Conservation of Angular Momentum'. But the Weather People call it the Coriolis Force, just to keep the name short — and I have no problem with

You spin me right round ...

The basics you need to understand

The spin axis is the shaded line. Parts of the body mass are further away from the rotation axis, which means a slower spin.

The spin axis is the shaded line. All of the body mass is close to the rotation axis, which means a faster spin.

The force that makes cyclones spin is the Coriolis Force. It's shown best via the skater above. The Coriolis Force is the driving force behind the spinning of cyclones.

The Coriolis Force is strongest at the North and South Poles and zero at the equator (because the Earth's surface at the equator is parallel to the Earth's spin axis).

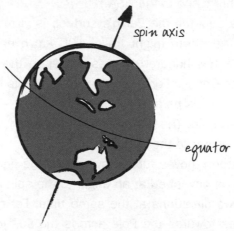

spin axis

equator

that. You can see angular momentum being conserved (or saved) when an ice-skater spins on one foot. When the skater pulls in their hands close to the centre (or spin axis) of their body, they speed up. And when they stretch out their arms, they slow down again. This is the driving force behind the spinning of cyclones.

The spin axis of an ice-skater runs from the top of the skull to between their feet. The spin axis of Earth runs through the North and South Poles.

The Coriolis Force is zero at the equator, because the Earth's surface at the equator is parallel to the Earth's spin axis. As you move over the Earth's surface at the equator, you stay the same distance from the Earth's spin axis.

The Coriolis Force is strongest at the North and South Poles because the Earth's surface is virtually perpendicular to the Earth's spin axis. For each 100 km that you move closer to the South Pole from the equator, you are probably moving 99 km closer to the spin axis.

Cyclone Moves (Fairly Hard, Okay to Skip)

Many times each year, a big storm will start up near the equator and then move towards either the North or South Pole. (By the way, this huge storm is called a cyclone in Australia, a typhoon in Asia and a hurricane in the USA.)

Near the equator, the Earth's surface is virtually parallel to its spin axis. The storm might travel 100 km over the surface, but move only a few hundred metres closer to the spin axis of the Earth, so the Coriolis Force is very small. This is like the ice-skater moving their hands parallel to their axis of spin — it doesn't affect how fast they spin. The Coriolis effect happens only when they pull their hands closer to the spin axis.

As the storm moves further away from the equator, the surface of the Earth is now tilted at an angle to the spin axis; the storm is moving in two directions at the same time. For every 100 km the storm moves towards the Pole across the surface, it also moves

Dis Information and Other Wikkid Myths

(say) 30 km closer to the spin axis of the Earth. This storm carries a few million tonnes of water, and it's spinning. A big storm like this is the military-industrial version of the ice skater pulling their arms in closer to the spin axis of their body.

I won't go into the maths of it but, to balance the momentum equations, the storm spins clockwise in the southern hemisphere and anticlockwise in the northern hemisphere as it moves away from the equator.

The Coriolis Force in War

Infantry units see the Coriolis Force in action in long-range guns. The operators in charge of fire control of long-range guns have to apply different corrections to the left or right, depending on how far from the equator, and in which hemisphere they are when firing the guns. These situations involve large distances — a cyclone can span hundreds of kilometres, while a really big gun can easily lob a shell 50 km. But a toilet or hand basin is much smaller — only about 30 cm across. And if you do the numbers, the Coriolis Force on such a small body of water is about 10 million times smaller than the gravity force of the Earth pulling on the water in the toilet or hand basin. So could you possibly see such a microscopic Coriolis effect when the 'gravity effect' is so much bigger?

The answer is 'yes' — but only if you are extremely careful. The rotation effect is so small that it is vastly overshadowed by local effects, including the direction in which the water entered the toilet bowl or basin, wind effects over the surface of the water, temperature effects, and so on.

What You See is What You Get

The experiment has been conducted in both hemispheres. It was first done in 1962 by Ascher H. Shapiro at MIT in Massachusetts, and then repeated in 1965 by Lloyd M. Trefethen at the University of Sydney.

Shapiro had a very shallow, round symmetrical dish about 2 m across and 150 mm deep. The outlet hole was about 9 mm across, and in the dead centre. He added the water through a hose, deliberately swirling it clockwise (the opposite of the expected draining direction). He then covered it with a plastic sheet and let the water stand for 24 hours, to reduce the initial rotation of the water. He placed a small floating cork on the water, and then released the stopper plug from below. The water took 20 minutes to drain out, with no visible rotation for the first 12–15 minutes. Then he could see the cork begin to spin anticlockwise — slowly at first, then gradually increasing to one rotation every four seconds by the end. The University of Sydney team used very similar apparatus, and got consistent clockwise rotations.

Shapiro wrote in *Nature*, 'When all the precautions described were taken, the vortex was invariably in the counter-clockwise direction.'

So if you lurch off the plane at Singapore (about 1° from the equator), rush to the bathroom, fill the oval hand basin with the off-centre drain hole with water and pull the plug ... don't expect to see the water spinning anticlockwise down the plughole.

But what about a really powerful golfer, hitting a long drive in a north–south direction, when they are a long way from the equator? Maybe ...

Foucault Pendulum

If you are very careful, you can see the Earth's rotation in the regular swinging of a pendulum. The pendulum will keep on swinging back and forth in the direction in which you started it moving, but the Earth will rotate underneath it, making it look as if the pendulum is gradually rotating.

A team consisting of Peter Maul, Rod Laing-Peach, Caroline Pegram and myself built a series of Foucault pendulums.

Back in 1848, Léon Foucault set up a long, skinny metal rod in his lathe. He accidentally 'twanged' it, and the end of the rod vibrated up and down. If you see the chuck of the lathe as a clock, the end vibrated from 12 o'clock down to 6 o'clock, and back to 12 o'clock, and so on. He slowly rotated the chuck by 90°. But the end of the metal rod steadfastly vibrated back and forth between 12 and 6 o'clock!

He then set up a 2-m-long pendulum with a 5 kg ball. The ball swung back and forth, and the line of the swing slowly rotated clockwise around its centre.

Imagine that you set up a Foucault pendulum at the geographic South Pole. You release the heavy ball. Over the day, the ball will keep on swinging in the same line in which you first launched it, relative to the distant, unmoving stars. But the Earth will rotate 'under' the Pendulum. So over a 24-hour day, the swing of a Foucault pendulum will appear to gradually sweep out a 360° circle.

Tricking Michael Palin

In his fascinating TV series *Pole to Pole*, actor and adventurer Michael Palin meets a man named Peter McLeary in Nantuki, Kenya, right on the equator. Peter gives a standard lecture to tourists in which he 'shows' that water goes down the drain in opposite directions, depending on which hemisphere of the Earth he is in. He says, 'This is the northern hemisphere [pointing to the left], and this is the southern hemisphere [pointing to the right]. If you drain a sink when you're on the northern side of the equator, and you watch the water as it drains, you will see that the water always rotates clockwise.' At this stage, he already has the direction of rotation wrong! But he holds in his hands a square dish half full of water which has a hole in the bottom that he blocks with his finger.

Peter walks a few metres into the northern hemisphere — and, sure enough, once he releases the water, the floating matchsticks on its surface are spinning clockwise. He then walks into the southern hemisphere. As he takes his finger off the drain hole, the water spins anticlockwise!

How does he do it?

First, when he wants to water to spin clockwise, he spins around to face the audience in a clockwise direction (and vice versa).

Second, he uses a square dish, which tends to 'lock' the water in the bowl better than a round dish. So when he spins clockwise, the water in the bowl is more likely to spin clockwise.

Anyhow, the position of the actual equator is a fuzzy thing, and certainly not defined to a metre or so. And when Peter walks only a few metres from the 'equator', he is still keeping the same distance from the spin axis of the Earth, because the surface of the Earth is parallel to the spin axis — so there would be hardly any Coriolis effect. Still, it makes good television …

References
Shapiro, Ascher H., 'Bath-Tub Vortex', *Nature*, Vol 196, 15 December 1962, pp 1080–1081.

Torok, Simon, 'Which way down the plug hole?', *The Helix*, December 1996/January 1997, p 25.

Trefethen, Lloyd M. et al, 'The Bath-Tub Vortex in the Southern Hemisphere', *Nature*, Vol 207, 4 September 1964, pp 1084–1085.

Gun Recoil – a Bulletproof Theory?

Movies are all about images. One indelible Hollywood action image is the Thrown-Back-by-a-Bullet one. In this sequence you see the victim being hit by a bullet and then lifted off their feet and thrown backwards. They usually land on a conveniently located large shopfront window which then smashes into a million pieces, in lovely slow motion. This special effect looks good, but it completely disregards the Laws of Physics and so creates yet another mythconception.

Movie Physics 101

To explain what should happen, you need a simple update on 'Conservation of Momentum'. To make it easy, let's assume that the victim and shooter weigh roughly the same. If the victim is going to be thrown backwards a certain distance, then the shooter has to be thrown back the same distance.

The bullet starts off in the barrel of the gun with zero velocity and zero momentum. The powder in the cartridge explodes and turns into gas; the gas expands and pushes the bullet ahead of it. Newton's Third Law tells us that 'For every action, there is an equal and opposite reaction'. Thus, the bullet zips away from the shooter and applies an equal force back on the shooter — the 'recoil'.

To keep things simple, let's assume that the bullet doesn't lose any speed in its flight because it's at such close range to the victim. It then hits the victim and comes to a complete halt inside the victim — which means that all the momentum of the bullet is now transferred to the victim.

Movie Physics 102

Now let's assume that the victim weighs 80 km. He's been hit with a bullet weighing around 30 g, travelling at around 1700 kph. This combination of mass and speed gives the bullet a certain amount of momentum.

If we ignore the losses due to air friction, and assume that all the bullet's momentum gets transferred to the 80 kg man, it turns out that the bullet gives the victim a velocity of around 0.7 kph! Now bear in mind that people can walk at about 6 kph, so 0.7 kph

Puttin' a cap in that theory!

When a gun is fired, the powder in the cartridge explodes and turns into gas. The gas expands and pushes the bullet ahead of it.

The 'zipping' bullet

The bullet zips away from the shooter and applies an equal force back on the shooter – the 'recoil'.

The highly exaggerated recoil

is just an accidental nudge. There's no way that a shooting victim is going to be flung off his feet and thrown into the plate glass window behind him with such a small nudge.

Going back to our speeding bullet: in reality it turns out that the victim gets less of a push than the shooter. First, the bullet does slow down in the air. Second, when the bullet hits the victim, not all the bullet's energy appears as a 'push' on the victim — some of it appears as a shockwave in their body, and some appears in the deformation/damage of the bullet and/or victim's body parts. So the bullet gives a smaller push to the victim than the shooter ... which is yet another reason why they would not get violently hurled into the nearest window.

Good Movie Physics

One movie that did show this action-and-reaction accurately was *Men in Black,* made back in 1997. Agent J (Will Smith) is assigned as an apprentice to the older and more experienced Agent K (Tommy Lee Jones). On his second day on the job, Agent J has to stop our planet from being destroyed. The agents go to an armoury and Agent K picks out a huge firearm, roughly half his body length, weighing half his body mass. He gives Agent J a tiny firearm not much bigger than a small butterfly, called the 'Noisy Cricket'. When Agent J fires the Noisy Cricket and destroys a garbage truck, he is sent flying backwards into a conveniently located pile of garbage bags.

Perhaps, though, there is reason why the shooting victim flies into the plate glass window, and it has nothing to do with the physics of the situation. Could I cynically suggest that when script, plot and continuity are not important to movie-makers, the Laws of Physics don't stand a chance ...

References

Men in Black, Columbia Pictures, 1997.
Intuitor Insultingly Stupid Movie Physics
http://intuitor.com/mpmain.html

Vitamin OD

Vitamins should be a pretty straightforward subject — the overall impression is that they are good for you. A logical extension is that if a little is good, then more must be better. The more of them you eat, the better off you will be — right?

Shelves in health food shops groan under the weight of vitamin pills and tonics. Australians spend about $800 million each year on 'complementary' medicines, and that includes a lot of vitamins. It's claimed that vitamins will help ward off cancer, ageing, hearing and vision loss, autoimmune disease — and possibly even sunstroke, syphilis and varicose veins. But this mythconception is dangerous, because too much of some vitamins can make you sick, bring on cancer — and possibly even kill you.

Vitamins

Vitamins are a family of organic chemicals that are essential for good health. They are different from 'regular foods' (such as fats, proteins and carbohydrates) in that they are not broken down to give energy. Instead, vitamins usually help regulate the chemical reactions of metabolism. Another difference between vitamins and regular foods is that you need only microscopic amounts of vitamins — usually between 0.00002% and 0.005% of your complete diet.

A vitamin deficiency is definitely bad. For example, a deficiency of vitamin A (usually derived from the livers of animals or fish, or from plants) can lead to night blindness, hardening of the skin, kidney stones and lung disease. A lack of Vitamin B1 (thiamine) can cause a whole bunch of problems, including beriberi (heart failure combined with dilated small blood vessels), Wernicke's encephalopathy (progressive mental deterioration, paralysed eye movements and disorientation) and other diseases of the nervous system.

A deficiency of niacin (nicotinic acid), another vitamin, causes pellagra, whose symptoms include dermatosis, diarrhoea and dementia. A lack of vitamin C (taken from fruit and vegetables) can make your gums unhealthy and your teeth fall out. A shortage of vitamin D (provided by fish and sunlight) can give you rickets (deformed legs).

Vitamins: the good, the bad and the ugly

It is generally thought that vitamin supplements are good, so therefore many (or megadoses) are great ...

The best line to adopt is to try to source them from their naturally occurring state ... called 'food'.

However, each person needs different amounts of vitamins under different conditions; for example, supplements of folic acid often given to pregnant women to prevent spina bifida in their unborn babies may not be recommended for the non-pregnant population. In general, recommended vitamin intake levels are set high enough to protect practically everybody, but not high enough to damage them. For example, the RDA (recommended daily allowance) of vitamin C is about 60–90 mg/day.

Megadose Therapy

It was Linus Pauling, the dual Nobel Prize winner, who popularised so-called 'megadose' vitamin therapy. He believed that huge quantities of vitamin C could help prevent or slow down the progress of heart disease, cancer and ageing. Back in the 1970s I read his book and was convinced enough to take 10 000 mg of vitamin C a day for about a year (that's 1000 times more than you need to prevent scurvy!).

Pauling also said, 'The proper intake of vitamin C helps keep one from catching colds.' Unfortunately, follow-up research has shown that megadoses of vitamin C do hardly anything for you. Pauling's belief that vitamic C improves life expectancy and protects against cancer and heart disease was never proved. Even a genius makes mistakes!

What megadoses of vitamin C can do, however, is reduce (by about 20%) the length of time that the cold makes you feel crook, especially if you take 8000 mg on the first day.

Thankfully, overdoses of vitamin C are relatively benign. Some people get diarrhoea, but back in my 10 000 mg/day phase, I had Guts of Steel.

You, however, may not be so lucky if you overdose on the B-group vitamin niacin — you could damage your liver. And overdoses of thiamine can cause respiratory failure and, sometimes, death.

Megadose Vitamin A

Dr Demetrius Albanes, a senior investigator in the Division of Cancer Epidemiology and Genetics at the US National Institute of Health, found a health problem with overdoses of vitamin A. When we eat chemicals called betacarotenes, our liver converts them into vitamin A. (Betacarotenes give carrots and pumpkins their orange colour.) Dr Albanes knew that previous studies showed that a diet (note: a diet, not pills) rich in betacarotenes protected people from lung cancer, so it seemed quite reasonable to assume that lots of purified betacarotenes would give greater protection.

In his study Dr Albanes enlisted 15 000 people, all smokers. He fed them (via vitamin pills) the betacarotene equivalent of six carrots each day. The study followed the 15 000 smokers for eight years but, just before it was due to end, they unfortunately had to pull the plug on it. Before they'd started, the first statistical analysis had predicted that many of the smoking volunteers would die from lung cancer if they didn't take the vitamin pills. But their early analysis of the results showed that an extra 20% of the study's volunteers had been diagnosed with lung cancer — in other words, the betacarotene was increasing their risk of lung cancer, not reducing it. That's something to think about the next time you buy a bunch of vitamin pills!

There are other risks from vitamin A, too — a separate Swedish study by Professor Melhus linked high vitamin A intake with a quite different disease, osteoporosis.

The Truth

Now, we all know that a bit of food is good for you — but we also know that too much food can be bad for you. In the same way, a bit of betacarotene (vitamin A) in your orange foods is good for you, but high levels in vitamin pills can be very bad for you. So, for most of us, nothing beats a balanced diet and a bit of exercise.

I wonder what Dr Albanes would have found if he could have convinced his 15 000 smokers to actually eat six carrots each day. And then there's Mae West, who wasn't thinking about food at all when she said, 'The only carrots that interest me are the number of carats in a diamond.'

History of Vitamin A

Chemists purified the chemical betacarotene from carrots back in 1910. But it took until 1950 before they could make it in the lab.

It was proved that vitamin A existed in 1913; chemists had worked out its chemical structure by 1933, but it took until 1947 to make it in the lab.

Vitamine to Vitamin

Back in 1911, the first vitamins were given the name 'vitamine', from the Latin word *vita*, meaning 'life', and the English word 'amine', referring to certain chemicals that contain nitrogen. In those early days, vitamins were identified only by a letter, e.g. vitamin A.

But once it was discovered that some of them did not contain nitrogen, and therefore did not contain an amine, the 'e' at the end of 'vitamine' was quietly dropped to give us the new word 'vitamin'. And once we worked out the chemical structure, the chemical name was added so that vitamin C, for example, was also called 'ascorbic acid'.

Vitamins in the Elderly

As we get older, and closer to death, our various systems run down. So many people have suggested that the aged should take vitamin pills. After all, they argue, vitamin pills are cheap and can do no harm. However, vitamin pills can chew up a lot of an old person's limited income — and, yes, vitamin pills can do harm.

Drs El-Kadiki and Sutton analysed eight randomised controlled trials that looked at vitamin supplements in the elderly and couldn't really see any single benefit emerging out of all the data.

So if you can afford it, and you avoid the pills that can harm you, it will probably hurt only your wallet, and it may help — but then again, it may not.

References

'Vitamin C left out in the cold' *New Scientist*, 2 July 2005, p 18.

El-Kadiki, A. and Sutton, A.J., 'Role of multivitamin supplements in preventing infections in elderly people: systematic review and meta-analysis of randomized controlled trials', *British Medical Journal*, 31 March 2005, pp 871–874.

Lee, et al., 'Vitamin E in the Primary Prevention of Cardiovascular Disease and Cancer: The Women's Health Study: A Randomized Controlled Trial', *Journal of the American Medical Association*, 6 July 2005, pp 56–65.

Shark Cartilage – the Great White Lie

Every now and then, a new 'wonder cure' for cancer comes along and quickly becomes incredibly popular. One current popular treatment is shark cartilage. After all, its proponents argue, 'Sharks don't get cancer, and sharks don't have bones, only cartilage, so it's obvious that shark cartilage prevents cancer.' If only it were that easy.

There are three problems with that deceptively simple sentence about sharks having no bones, and therefore not getting cancer.

Sharks Do Get Cancer

First, it is absolutely true that sharks do not have a skeleton made from bones. Yes, their skeleton is made from cartilage. But sharks do get cancer.

There are at least 42 cases on record of sharks having various tumours, including thyroid cancer, lymphoma, metastatic adeno-carcinoma, and other cancers of the nervous, blood, reproductive,

Dis Information and Other Wikkid Myths

skin and digestive systems. A few sharks had two tumours. And there was even a case of a tumour of the cartilage. This work was done by Dr Gary Ostrander, Professor of Biology at Johns Hopkins University, who surveyed the tissue database at the National Cancer Institute's Registry of Tumors in Lower Animals held at George Washington University. You might correctly say that 42 is not a huge number, considering how many sharks there are. And that's true, but most commercial fishers don't carefully examine their caught sharks for cancers. Anyhow, in the ocean most sick animals are eaten by healthy ones, or just sink to the bottom and die.

Cartilage Doesn't Work (and Stinks)

Second, all the studies done so far show that eating a cup or two of shark cartilage each day does not harm you. But not one of them shows that shark cartilage can prevent or treat cancer.

One study that looked at people with incurable colorectal or breast cancer was typical: over half the volunteers refused to take their very smelly and disagreeable shark cartilage after dutifully

Shark cartilage: the 'wonder cure'

Every now and then a 'wonder cure' for cancer comes along. A current popular treatment is shark cartilage. Proponents argue, 'Sharks don't get cancer, and sharks don't have bones, so it's obvious that shark cartilage prevents cancer.'

The truth: Sharks do get cancer. There are at least 42 cases on record of sharks having various tumours.

consuming it for a month, because one side effect of it was that they smelt bad.

Avoiding Regular Treatment

The third problem with shark cartilage is very serious: cancer sufferers have been convinced by advertising to buy shark cartilage and stop taking proven anti-cancer treatments. The anti-cancer treatments we have now are not perfect — but they are the best we have.

So how did we end up with the situation where shark cartilage is sold all over the Internet, and in health food shops, as a universal cure, vitamin or body-building aid? One particular preparation simply called 'TerraVita Shark Cartilage' claimed that it would treat 'arthritis, joint pain and health, osteoarthritis, stiffness and more' and that it 'Cures/Prevents Cancer, Promotes Wound Healing, and Relieves Arthritis Pain and Stiffness'.

Shark Cartilage History

The whole sorry shark cartilage saga began with I. William Lane in 1983. In the 1990s, he co-wrote two books with Linda Comac arguing that 'sharks don't get cancer'. His 1992 book was called *Sharks Don't Get Cancer: How Shark Cartilage Could Save Your Life*. The 1996 book was called *Sharks Still Don't Get Cancer*. He got lots of welcome publicity on the US version of *60 Minutes*.

This made it easy for his son, Andrew Lane, to set up Lane Labs in New Jersey to market and sell shark cartilage. One of their products that sold very well was called BeneFin — until the US Food & Drug Administration took out successful injunctions against Lane Labs for 'deceptive marketing' in 2004. Not only did Lane Labs have to stop selling BeneFin (and SkinAnswer and MGN-3), they had to pay a US$1 million fine.

Sharks and Anti-Cancer

Putting all of that to one side, there is something anti-cancer going on somewhere in sharks.

First, Dr Carl A. Luer, from the Mote Marine Laboratory in Sarasota, Florida, has been trying to give sharks cancer for the last decade, to learn more about human cancers. He finds sharks the hardest animal to give a cancer to. For example, the carcinogenic chemical Aflatoxin B1 (sometimes found in peanuts) quickly brings on cancers in fishes with bony skeletons, but it never induces cancer in sharks (with their cartilaginous skeleton). Does the cartilaginous skeleton have anything to do with this? Dr Luer doesn't know — and, at this stage, neither does anybody else. Maybe it's because sharks eat so many fish ...

Second, each cancer has an essential need to grow new blood vessels to feed it. There is something in fresh shark cartilage that can, in laboratory tests, slow down this growth of new blood vessels. But none of the chemicals that have been isolated have improved the life expectancy of cancer patients in all the studies done so far, even though they work on blood vessels in laboratory glassware on a laboratory bench.

And please note the 'fresh' in 'fresh shark cartilage' labelling: most of the sharks caught for this market are stored as dead carcasses, without refrigeration, often for days or weeks.

What A Waste ...

There is, however, another aspect to this story — the sharks themselves. In much of the ocean, shark populations are down to a few per cent of their levels a century ago. Why kill sharks, and remove an essential top predator, just to market worthless treatments that impoverish sufferers and set them up to not use treatments that can work?

Shark Teeth

In its lifetime, a shark grows somewhere between 10 000 and 30 000 teeth. We humans only have two sets of teeth in our whole lifetime, but sharks get new sets of teeth between every five months and once a week, depending on what sort of shark they are. These teeth can exert a biting pressure of three tonnes per square centimetre, and have been known to bite through high-tensile steel cables. Actually, they are not real teeth like regular animal teeth — shark teeth are more like modified scales. Nevertheless, they still do a pretty good job.

References
'Sharks DO Get Cancer', Random Samples, *Science*, 14 April 2000, p 259.

Baum, Julia K., et al., 'Collapse and Conservation of Shark Populations in the Northwest Atlantic', *Science*, 17 January 2003, pp 389–392.

Ostrander, Gary K., et al., 'Shark Cartilage, Cancer and the Growing Threat of Pseudoscience', *Cancer Research*, 1 December 2004, pp 8485–8491.

The Reasons
for Seasons

It has taken astronomers thousands of years to give us an accurate picture of the universe around us. But it seems that some of us haven't been paying attention.

In 1988, Matthew Schneps of the Science Media Group at the Harvard-Smithsonian Center for Astrophysics wrote, produced and directed an 18-minute film documentary, *A Private Universe*. In it, students, faculty and graduates of Harvard University were asked the seemingly simple question, 'What causes the seasons?' Twenty-three people were asked — only two got the right answer. I'm guessing that, even in the early 21st century, most people don't know the reasons for the seasons.

Most of the wrong answers fell into the category of 'If I'm closer to the fire, then I feel hotter'.

Distance From the Sun — No

It is true that as the Earth orbits the Sun it does indeed trace out an orbit that is not circular. The orbit is very slightly oval or egg-shaped. Over the course of a year, the distance between the Earth and Sun varies between 147 million and 150 million kilometres. Around 3 January, we are at our closest to the Sun, and about six months later we are at our furthest from the Sun.

There are three problems with the 'closer means hotter' answer. First, in the USA (where the questions were asked) 3 January corresponds to the dead of winter, not summer. How can 'closest to the Sun' cause winter? Second, why are the seasons different by six months on each side of the equator? Why isn't it summer in both northern and southern hemispheres at the same time?

Third, that change in distance (less than 2%) would change the temperature of the Earth by about 4°C: that is much less than the temperature change that occurs as we ceaselessly swing from summer to winter and back again.

Tilt of the Earth — Yes

The reason for the seasons is the tilt of the spin axis of the Earth.

All the planets orbit around the Sun in roughly the same horizontal plane. If the North Pole-South Pole spin axis of the Earth was at 90° (exactly upright) to this plane, the amount of daylight we experience would be 12 hours everywhere on the planet. (This ignores the 'bending' effect of the thicker atmosphere at the horizon at sunrise and sunset, which gives us a few extra minutes each day.)

But the north-south spin axis of the Earth is not at 90° to this horizontal plane: it's tilted about 23.5° from the upright position. And it maintains this tilt year in and year out as it orbits the Sun.

This tilt has two effects: first, it controls how many hours of sunlight fall on different parts of the planet; second, it controls how 'effective' this sunlight is.

Hours of Sunlight — Yes

Around 22 December each year, an observer sitting on the Sun would see much more of the southern hemisphere than the

It's the season for it

The seasons depend on the tilt of the Earth.
This then controls how many hours of sunshine we
get each day, and how intense that sunshine is.

The seasons on Earth

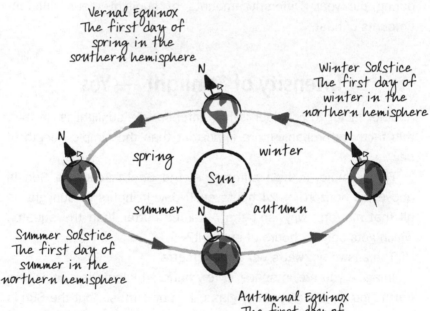

Vernal Equinox
The first day of
spring in the
southern hemisphere

Winter Solstice
The first day of
winter in the
northern hemisphere

spring

winter

Sun

summer

autumn

Summer Solstice
The first day of
summer in the
northern hemisphere

Autumnal Equinox
The first day of
autumn in the
southern hemisphere

23.5 degrees

Axis of
Earth's
rotation

northern hemisphere as the Earth spins on its axis every 24 hours (they'd also need extremely good air conditioning!). Sydney (at 34° from the equator) would get about 14.5 hours of sunlight at this time, and would be enjoying summer.

About three months later, our observer on the Sun would see roughly equal amounts of each hemisphere as the Earth spins.

And after another three months, our observer would see much less of the southern hemisphere, and Sydney would get only 9.5 hours of sunshine in its winter. (The difference in the hours of sunshine is zero at the equator, and greatest at the Poles.)

Yes, the Earth is tilted, so we get different amounts of sunlight during the year. Different amounts of sunlight mean different amounts of heat.

Intensity of Sunlight — Yes

Another factor to consider is how intense this sunlight is. In fact, this factor is probably more important than the simple 'length of day'.

For example, in high summer at the South Pole, the Sun is above the horizon for 24 hours each day. If 'hours of sunlight' is all that matters, why isn't the Antarctic hotter than the equator, which gets only 12 hours of sunlight?

The answer is 'watts per square metre'.

Imagine you are in space, a few hundred kilometres above the Earth. You lay out a sheet of glass, 1 m by 1 m, so that the Sun is directly above it (not tilted). One square metre of sunlight falls through it and plunges down towards the Earth.

If you are directly above the equator, that 1 m^2 of sunlight lands on 1 m^2 of surface, and dumps about 1000 watts of power into the ground. That's a lot of heating.

From the ground you can see the Sun directly above you. But if you are above the South Pole, that 1 m^2 of sunlight is spread across 10 m^2, because of the curve of the Earth. So each square metre might get only 100 watts of power — not 1000 watts —

therefore the heating is a lot less. The sunlight is hitting the ground at an oblique angle, so you get less power delivered per unit area. And from the ground you see that the Sun is barely above the horizon.

During the Antarctic summer, the Sun is close to the horizon, so not a lot of heat is dumped on each square metre of the ice, and the temperature stays sub-zero. (Another factor is that the ice is very reflective.)

This extra factor (fewer watts per square metre) makes the Poles a lot colder than the equator.

The Reasons for Seasons

So the seasons depend on the tilt of the Earth, which then controls how many hours of sunshine you get each day, and how intense that sunshine is.

That knowledge may not make us wealthy — but it may make us feel more at home in our enormous universe …

References

Domins, Neil F., *Heavenly Errors: Misconceptions About the Real Nature of the Universe*, Columbia University Press, 2001, pp 5–6, 12–19, 103, 112.

Schneps, Sadler, et al., *A Private Universe*, produced by the Harvard-Smithsonian Center for Astrophysics, 1987, ISBN 1576804046.

Tighten the Asteroid Belt

Whenever the plot slows down in a sci-fi movie, the scriptwriter will throw in the dreaded Asteroid Storm or Belt. For example, in *The Empire Strikes Back* the good guys in the *Millennium Falcon* fearlessly plunge into an Asteroid Belt to get away from the bad guys. Of course, the baddies (being bad) are not very good pilots and smash into the rocks, while the goodies (who are, of course, better pilots) get through alive. But it was a close shave, even for the goodies. The rocks are huge, there are thousands of them, and they are so close together that the good guys have to dodge, weave and duck ceaselessly to avoid collision. Indeed, the rocks are so close that you could sometimes jump from one to the next. In fact, the very name — Asteroid Belt — suggests a continuous body. But in reality, it's nothing like that. It's practically all empty space.

Asteroid Belt

In our solar system we have a well-known asteroid belt orbiting the Sun. It lies between the four inner small, rocky planets (Mercury,

Venus, Earth and Mars) and the four outer giant, gas planets (Jupiter, Saturn, Uranus and Neptune). Most people have vaguely heard of this asteroid belt.

It took a while for our telescope technology to get good enough, but on 1 January 1801, an asteroid was discovered in this belt. It was also the largest ever discovered — about 930 km across. Of course, the smaller asteroids in this belt took longer to discover because they were harder to see. As the astronomers discovered tens, then hundreds and, finally, thousands of asteroids, the name 'asteroid belt' gradually evolved.

This asteroid belt has about one million rocks bigger than 1 km across, and about 250 of these are larger than 100 km across. There are many millions more rocks, just a few metres across. Even so, the total mass of all the asteroids is about 1000 times smaller than the mass of the Earth. The current theory is that the asteroids are the remnants of a planet that couldn't quite coalesce into existence, probably because of the disruptive effects of the enormous gravitational field of Jupiter.

However, it appears now that there are two more zones where asteroids hang out.

The universe is a big place, and distances are huge. Astronomers call the distance between the Sun and Earth (about 150 million kilometres) one astronomical unit (AU). Jupiter (the first of the giant gas planets) is about 5 AU from the Sun, while Neptune (the last of the giant gas planets) is about 30 AU out.

The last planet, Pluto, is a small, rocky planet. It has a very elliptical orbit that takes it between 30 and 50 AU from the Sun. It cannot hit Neptune, because the two planets are locked into mathematical 3:2 gravitational resonance. This means that in the time it takes Neptune to make three orbits of the Sun (about 165 years per orbit), Pluto will make two orbits (about 248 years per orbit). They are locked into this, and never come close enough to hit.

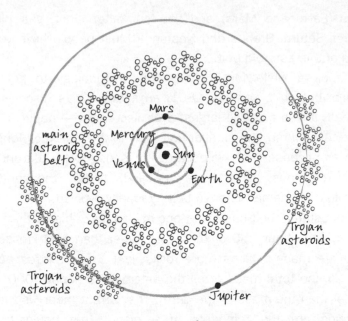

In our solar system we have a few asteroid belts orbiting the Sun. The best-known one lies between the four inner small rocky planets (Mercury, Venus, Earth and Mars) and the four giant gas planets (Jupiter, Saturn, Uranus and Neptune).

Kuiper Belt

There is another asteroid belt past Pluto called the Kuiper Belt.

Back in 1951, the Dutch-born American astronomer Gerard P. Kuiper was thinking about comets, especially those that loop around the Sun in less than 200 years. These comets include Halley's Comet (last seen in 1985/86) and Shoemaker-Levy (which slammed into Jupiter in 1994). He worked out that comets like these would have to come from a belt fairly close to the planets of the solar system. He also noticed that these comets tend to come hurtling in from outer space quite close to the plane of the planets, rather than from any old direction. Kuiper predicted that there should be a flattened belt or disc of comets and asteroids beginning just outside the orbit of Neptune (30 AU) and reaching out to about 1000 AU. But finding a comet past Neptune is like trying to see a 100-watt light bulb at 20 times the distance of the Moon.

Dis Information and Other Wikkid Myths

In 1992, our telescope technology finally got good enough. After five years of looking, two astronomers found a comet in what is now called the Kuiper Belt. It was a biggie, too — about 280 km across. Since then, ground telescopes have found hundreds of big comets (between 100 and 400 km across), while the Hubble Space Telescope has found hundreds of small comets (a few kilometres in diameter). There are probably a billion comets and rocks in the Kuiper Belt.

In fact, in August 2005 astronomers announced that they had found something roughly the size of Pluto in the Kuiper Belt.

Oort Cloud

We think that another place where comets and rocks hang out is the Oort Cloud, named after Jan Hendrik Oort. This enormous, spherical cloud is a long way out at about 10 000 to 100 000 AU from the Sun, or approximately one-fifth of the way to the nearest star. We haven't actually seen comets in the Oort Cloud with our telescopes, but mathematicians tell us it has to be the home of the comets that come into the inner solar system from any direction (rather than in the plane of the planets). These comets take a long time to fall in, and then they take a long time to come back and visit us again. We think that Hale-Bopp (which swung past in 1997, and is about 40 km across) is one of these comets.

Asteroid Collision?

Back in the late 1960s and early '70s, NASA scientists and engineers were building a spacecraft to go deep into the solar system. It was Pioneer 10, a 260 kg, 3-m diameter vehicle. They launched it on 2 March 1972 to explore Jupiter and beyond — and it had to cross the asteroid belt.

Would it survive intact, or would it be smashed into pieces as soon as it entered the belt? Pioneer 10 carried no technology that could find or avoid asteroids, so the scientists kept its twin,

Pioneer 11, in storage and waited anxiously for seven months until Pioneer 10 successfully emerged untouched. The scientists then launched Pioneer 11 on a mission to Saturn and beyond — and it, too, passed through the asteroid belt unscathed. Since then, many spacecraft have successfully travelled through the belt.

It turns out that the asteroids in the belt are many millions of kilometres apart. In fact, the asteroid belt is so sparsely populated that whenever they plan to send a spacecraft through it, the scientists go to a lot of trouble to send it on a path close enough to an asteroid to take a look.

Tempel 1

A good example of this kind of space exploration is the bull's eye that NASA's 'Deep Impact' spacecraft made onto the comet Tempel 1. It's a moderate-sized comet, about 11 km long and 6 km across, and takes about 5.5 years to orbit the Sun, with an orbit swinging between Mars and Jupiter. The impact was timed to occur when Tempel 1 was at its closest to the Sun.

The two-part spacecraft had been launched six months earlier and travelled 134 million kilometres to get to exactly the right point. About 24 hours before the actual impact, springs pushed apart the two parts of the spacecraft. One part was the 372 kg copper 'impactor', about the size of a washing machine. It was going to hit the comet at a speed of some 37 000 kph, delivering as much energy as exploding 4.5 tonnes of dynamite would. The other part was the mothership, which would film the encounter and radio back any data from the impactor.

For the last two hours of its journey, the impactor spacecraft was on its own, making all of its own decisions — this was because the transit time for the radio signals to travel to and from the spacecraft was so long. The impactor had to make three separate course corrections to make sure that not only would it hit the comet, but that it would do so on an area lit by the Sun (so that the various cameras could get good pictures).

The impactor scored a direct hit at 0625 GMT on 4 July 2005. It took the combined work of hundreds of scientists and engineers to aim it, and it produced a crater the size of a city block, plus a huge icy plume.

This impact was also photographed by another comet chaser, Rosetta, from a distance of 80 million kilometres. Rosetta is also on a collision course with a comet. It will hit its target, Churyumov Gerasimenko, in 2014.

Space is very big, and rocks are very small in proportion. The images of the Goodies weaving their spaceship through the closely spaced free-flying rocks are pure fantasy — but lots of fun to watch.

Leaky Asteroid and Comet Belt

A comet is a ball of ice and dust. Halley's Comet is about 15 km long, while Hale-Bopp is about 40 km across.

Now, here's one *big* mystery about comets …

Halley's Comet swings by every 76 years or so. Each time it zips past the Sun, it flares up and throws away, as a beautiful tail, about *one ten-thousandth* of its weight. In other words, it should survive only *10 000 orbits* — maybe 750 000 years. But according to our best theories of the Origin of the Solar System, it seems that comets were made when the Sun and the rest of the Solar System popped into existence, about 4600 *million* years ago.

So where has Halley's Comet been hiding out? Almost certainly in the Kuiper Belt. And, presumably, other potential comets are hiding out in the Oort Cloud.

But just how do these comets get bumped out of their cosy deep-freezes?

One theory is that, every few million years, a distant passing star disturbs the Oort Cloud with its massive gravity. This would start a few comets on a million-year trip to the inner solar system. That's an okay theory for the Oort Cloud, which is a long way out, but what about the Kuiper Belt, which is a lot closer?

A recent theory is the Kuiper Belt has a 'slow leak' at its inner edge, just near Neptune. The gravitational field of Neptune is strong enough to actually disturb the Kuiper Belt, and so a comet is pulled towards Neptune and then gravitationally passed on, in turn, to Uranus, Saturn and Jupiter, and finally tossed towards the Sun. According to this theory, a comet from the Kuiper Belt should hit Jupiter every 400 years, and Earth every 13 million years or so (which would make an awful mess).

So for the continued survival of life on our planet, you'd better hope that Neptune doesn't rock the Kuiper Belt *too much*.

8 or 10 or 11 planets?

In October 2003, three astronomers — Mike Brown, Chad Trujillo and David Rabinowitz — photographed 2003 UB313, a chunk of rock and ice orbiting the Sun about 97 AU out — way past Pluto, but still in the Kuiper Belt. In January 2005 they reanalysed their data and realised that this lump was big — about 2300–3200 km across, which is a little bigger than Pluto, at 2390 km across.

Also in 2005, astronomers from the Sierra Nevada Observatory in Spain discovered yet another big lump. Object 2003 EL61 is also in the Kuiper Belt, and it's about two-thirds the size of Pluto and about 51 AU from the Sun.

The problem is this: not all astronomers are convinced that we should call Pluto a planet. This is because Pluto is too small; it's the smallest 'planet' in the solar system. It's significantly smaller than both Mercury (4879 km across, the innermost planet) and Titan (5150 km across, the giant moon of Saturn). These astronomers reckon that it should not have been included as a planet when it was discovered back in 1930.

If we include Pluto as a planet, then we have to include 2003 UB313 — so then we will have ten planets in the solar system. Do we then have to include all other objects as big as Pluto that we find in years to come?

And where is the cut-off point? Do we include Object 2003 EL61, which is two-thirds the size of Pluto?

In future years, the astronomers will decide. If they dump Pluto, then we will have eight planets in the solar system. If they keep Pluto, we will have at least ten planets …

References

'Hubble Confirms Kuiper Belt', *Astronomy Now*, August 1995, p 7.

Comins, Neil F., *Heavenly Errors: Misconceptions About the Real Nature of the Universe*, Columbia University Press, 2001, pp 5–6.

Leary, Warren E., 'Spacecraft Hits Passing Comet, Just as Planned', *New York Times*, 5 July 2005.

Theokas, Andrew, 'The Origin of Comets', *New Scientist*, no. 1599, 11 February 1997, pp 42–45.

Hit and Myth

Like most people, you probably believe that torpedos sink ships by setting off a big pile of high explosives as they bang into them. Back in 1999, the Australian Navy's submariners hadn't fired a live torpedo for over a decade (they can cost over a million dollars each). So the submariners were delighted when, on 14 June, the Navy had the new Collins Class submarine, HMAS *Farncomb*, sink a decommissioned warship, *Torrens*, with a Mk48 torpedo. But they weren't so delighted when the media reported that the torpedo hit the *Torrens* — because modern torpedos don't kill their targets by 'hitting' them.

'Damn the Torpedos ...'

Naval warfare has advanced a long way since Admiral Farragut (1801–1870) declared, 'Damn the torpedoes, full speed ahead.'

David Farragut went to sea as a youth. The ship he served on, the frigate *Essex*, captured so many British whaling ships in the War of 1812 that he was put in charge of one at the age of 12. Farragut also had many outstanding naval victories during the American Civil War (1861–1865).

Back then, the word 'torpedo' meant 'a floating mine' or 'an explosive device for detonation underwater'. On 5 August 1864,

The bang down below

Spherical gas bubble ——— Shock wave

The torpedo explodes under the keel of the ship.
An extremely high-pressure shock wave is generated,
moving through the water which begins to
load the hull.

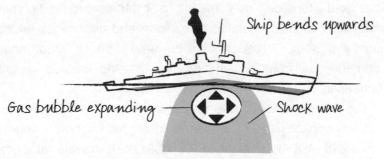

Ship bends upwards

Gas bubble expanding ——— Shock wave

The shock wave crushes and accelerates the underside of
the ship's hull, causing the ship to bend and fracture.
The expanding gas bubble then pushes the hull upwards,
after the shock wave has passed.

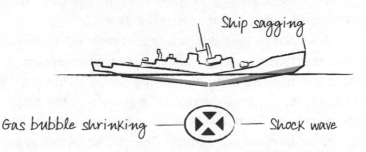

Ship sagging

Gas bubble shrinking ——— Shock wave

The gas bubble then shrinks, causing the ship to 'sag'.
This is called the 'whipping' phase and is the reason why
the torpedo is detonated underneath the ship.

Farragut's fleet entered Mobile Bay, Alabama, in two columns of ships. The *Tecumseh*, the lead ship of one of the columns, was destroyed by a mine. All the ships slowed down and began to drift confusedly — but they were still under the enemy guns of Fort Morgan, on the edge of the bay. Something had to be done immediately, or else the fleet would be destroyed. So Farragut shouted out his now-famous line, manoeuvred his ship, the *Hartford*, to the front and charged through the mine field — and none of the mines exploded.

Modern Torpedos

The word 'torpedo' now means a cigar-shaped self-propelled underwater missile intended to explode and destroy its target — usually a ship. Torpedos can be fired from a ship, another submarine, or from a plane or missile. One method classifies them as light-, medium- and heavyweight.

The American Mk48 is a very capable heavyweight torpedo. It first saw service in 1972, and in its latest incarnation is known as the Mk48 ADCAP (Advanced Capability). It travels at highway speeds, and has a range of 38 km at 55 knots (102 kph), or 50 km at 40 knots (74 kph). It weighs around 1600 kg (depending on the version), of which 267 kg are explosive, with the punch of 544 kg of TNT. It is in service in the navies of the USA, Australia, Israel, Canada, Turkey and the Netherlands.

The Mk48 can be guided by a wire that reels out behind it, or it can find the target by itself. Once it gets close enough to the target, it first uses sonar to aim for the centre of the ship. When it's really close, it uses the magnetic signature of the target as a trigger to explode, when it's about 15 m directly under its hull; the depth and location are quite critical. The 267 kg of high explosives almost instantaneously all turn into gas.

Amazing Physics

Then things get interesting (from a physics point of view).

First, the actual explosion (which converts a small volume of solid into a huge volume of gas) generates a very high-pressure shock wave. This rams into the middle of the underside of the ship's hull at about 1.5 km per second.

Second, the shock wave crushes the underside of the hull, and also lifts it up. It causes the ship to bend upwards in the middle, like a banana. A huge puff of smoke comes out of the target's smoke stack, thanks to the sudden upward acceleration from the shock wave; the upper decks of the ship crack apart. After a few hundredths of a second, the shock wave has come and gone. But within a few more fractions of a second, the expanding bubble of gas from the explosion hits the bottom of the hull. The bubble reaches a maximum size of about 18 m across, and it maintains the massive upward force on the bottom of the hull once the shock wave has passed. So the ship is bent upwards in the middle in two stages — from the shock wave and then the expanding gas.

Third, after about half a second, the bubble (thanks to some fancy physics) begins to shrink. The ship then 'sags' in the middle, and begins to banana in the other direction. This breaks up the hull of the ship even more. Navy people call this sagging the 'whipping' phase. It's actually very useful in breaking the back of a ship — after all, if you want to break a stick it's much more effective to bend it back and forth rather than in only one direction.

Fourth, after about one second the shrinking bubble has reached its minimum size and begins to expand again. The water pressure around it is greatest directly underneath (being further from the surface) and least at the top (being closest to the surface). So it tends to expand upwards more than downwards, and it pushes a lot of water upwards as a high-speed jet. This bananas the ship back in the first direction, and at the same time a second plume of smoke squirts out of the smoke stack.

Finally, the water jet and the enormous bubble ram through the hull. Combined, they can be powerful enough to completely rip the superstructure clean off the ship, giving the appearance of a second explosion. The ship is often snapped into two separate halves as a result.

On 14 June 1999, the submarine HMAS *Farncomb* fired its wire-guided Mk48 torpedo at the 2750-tonne decommissioned warship *Torrens* from over the horizon. The two 'explosions' happened 1.3 seconds apart — two hits for the price of one. The plume of water and ship fragments shot 150 m into the air; the *Torrens* split into two halves. The stern sank first, but the bow floated for a few hours before it too sank.

Please note that there is no red flame in any of this in real life. But there will be in a Hollywood movie, because they can't help themselves and will add it in.

Straight Physics

But it would all be very different if the torpedo slammed into the hull of the ship and then exploded.

A ship in battle mode makes all its separate compartments watertight by shutting and locking all the doors. The damage would be immense in a few compartments, which would get flooded. But the ship would probably keep on floating, with a severe list to one side. The remaining crew could be evacuated, and the ship could even possibly be repaired.

By the way, most modern warships are not heavily armoured; they can be destroyed by a single missile, so it's not worth it to armour them.

The Truth

So while a missile kills a ship by going *bang* on the body of the ship, a modern torpedo kills a ship by going *bang* underneath it.

Underwater Gun?

It seems ridiculous to even think about the concept of an underwater gun. Yes, the explosive charge would fire, it would turn into gas, and then it would try to push the bullet or shell against water. Water weighs one tonne for each cubic metre and provides tremendous resistance, so the bullet or shell would come to a stop within a few metres.

But what if you enclose the bullet in a bubble of gas and keep on generating this gas in front of the bullet? The bullet would never even know that it was in water — it would see only gas. For example, the water near the tip of the bullet might be vapourised by the speed of the bullet, and turn into steam — which is a gas. So it would 'fly' underwater. The technique is called 'supercavitation'.

The US Office of Naval Research is working on the Advanced Underwater High Speed Munition. One version is a 6.25-inch (15.875-cm) diameter self-protection weapon. These underwater bullets have been successfully fired from an underwater gun at speeds greater than the speed of sound in water (1.5 km/second).

They could protect a submarine or ship from torpedos by ramming the torpedos. They are being developed for the US Navy as the Rapid Airborne Mine Clearance System, which involves shooting bullets at underwater mines from airborne helicopters.

Reference
Encyclopaedia Britannica, Ultimate Reference Suite DVD, 2005.

Acknowledgments

Big it up for the peeps from the Doc's 'hood (see pix).

This book would not have been possible without the dedication and hard work of a few people in particular, so thanks and a Big Shout Out to:

Sophie 'The Hatchett' Hamley, Ali 'The Axe' Urquhart, Adam 'Scissors' Yazxhi, Caroline 'Pegs' Pegram, Lesley McFadzean, Dan the Man Driscoll, Neil Walshe, Jane Burridge, Shona Martyn, Louise Cornegé, Judi Rowe, Graeme Jones, Tracey Gibson, Katy Wright, and of course to my beautiful family.

Clockwise from top left: Deano Dave Day, Team Lab, Main Man Phil D, Geeks & Lil' Mel M, Sal-Lo, Peep VC of Sci Prof Beryl, Home Crew with 'Tude, Jazzy Jas & Wiz Merlin, Nano Nano Robyn W, Mel & Kochie ... Word.

Clockwise from top left: Scrubber Mel B, Office
Ho's, Yo Kadee & Ian Allen, Roger M Wikkid, Lil'
Literary Lesley, Hatchett & Axe, Hands-on Willis &
Leftie Lee, Adrock, Team Claxton.

Other Dr Karl titles …

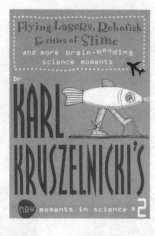

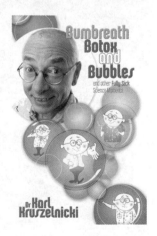